Principles of Fire Safety Engineering

Principles of

FIRE SAFETY ENGINEERING

Understanding Fire and Fire Protection

Second Edition

AKHIL KUMAR DAS

BTech (ChE), MS(FPE), DM, FIE(I), FIFE(UK)

Visiting Professor, University of Petroleum & Energy Studies, Dehradun
Former Head of Fire Safety Engineering & Management, ICEM Oman
Chief Manager, Chief Safety Officer & Management Representative ISO 14001 EMS, BRPL
Deputy Manager (Technical), Loss Prevention Association of India Ltd.

PHI Learning Private Limited

Delhi-110092
2026

In fond memory of ***Shri Asoke K. Ghosh*** *(October 1942 – February 2024), Founder Chairman and Managing Director of PHI Learning, whose vision endlessly inspires.*

The Legacy Continues....

Published by Pushpita Ghosh, PHI Learning Private Limited, Rimjhim House, 111, Patparganj Industrial Estate, Delhi-110092 and Gopsons Printers Pvt. Ltd., A-14, Sector-60, Noida, Gautambudh Nagar, Noida, Uttar Pradesh-201301.

₹595.00

PRINCIPLES OF FIRE SAFETY ENGINEERING: Understanding Fire and Fire Protection
Second Edition
Akhil Kumar Das

ISBN-978-93-89347-23-4 (Print Book)
ISBN-978-93-89347-24-1 (e-Book)

The export rights of the book are vested solely with the publisher.

To

the "Students" of Fire Safety

Contents

Preface

At the outset of this preface written for the second edition of *Principles of Fire Safety Engineering—Understanding Fire and Fire Protection* I wish to thank all the readers who have used this book as text or reference. The demand for the book since its publication in 2014 points to a lacuna that existed and I am happy to have filled this space. The feedback I have received from my students and mentors, and the reviews I have seen on e-commerce sites have further motivated me to bring out the new edition.

I am, however, a firm believer of the saying "there is always scope to improve" and have tried to do that in this new edition by adding a new chapter introducing readers with the very important subject of Modelling for Fire Safety, updating some statistics and making a few minor changes.

I realise that like all other fields, *Fire Safety Engineering* too is growing, changing and getting more popular. A greater number of new professionals are also entering the field as are mid-career entrants. There is thus a growing demand for books on the subject for students and others who want to learn and get back to a source. For them, a text is irreplaceable. My book is a step in that direction.

Your constructive feedback is always welcome.

Akhil Kumar Das

Preface to the First Edition

"The fire you kindle for your enemy often kills yourself."

—Anon

The study of combustion chemistry and the use of heat generated from combustion has been central to human development. Fires that went out of control, however, were generally analysed only after they had done damage. It was insurance companies that sponsored some of the earlier research into fires and fire protection. Out of such works were born multidisciplinary subjects like fire science and fire protection bringing together knowledge of a wide range of subjects like chemistry, applied mechanics, and civil, chemical, electrical and structural engineering, architecture, electronics and even psychology. Specialists from these wide-ranging fields studied different aspects of fire and saw fire in the same way as the six blind men "saw" an elephant. Some entrepreneurial minds brought together these specialist perceptions and gave birth to a new subject like that of fire dynamics which is derived from areas as varied as chemistry, thermodynamics and aerodynamics. The field of fire protection draws on fire dynamics and combines it with the knowledge of subjects like hydraulics and electronics.

Two handbooks published by the National Fire Protection Association (NFPA) and by NFPA in association with the Society of Fire Protection Engineers (SFPE), authored by experts in their respective fields, have brought together a wide canvas of related subjects. Numerous books have been written by the authors with focus on their specific areas of interest and experience. "Students" of fire, however, continue to be constrained by the unavailability of fundamental information about fire at one place. I have observed this more closely while teaching formal degree programmes in HSE and Fire Engineering. The dearth of literature is more pronounced in India, where study or research on the subject has been confined to a miniscule community of specialists.

As the first Indian (and among the first worldwide) to earn a formal postgraduate degree in fire protection engineering and having worked in the profession for over three decades in industry, consulting, teaching and training, I feel a responsibility to put together the fundamental aspects of fire science and fire protection in the form of a book and hence the present text. While working on the manuscript, I found it difficult to precisely identify the scope of coverage and have tried to include the aspects of theory that aid the understanding of practice. Though the syllabus of several programmes of study in the subjects of fire and safety were referred, I have extended the scope by including those areas I perceive should also be included in the curricula.

To find an Indian publisher for the book was not an easy task. I am thankful to the team at PHI Learning who saw the space for a textbook on the subject suited to formal academic curricula in the country.

This book is not intended to be either a handbook of all subjects that make up this field nor does it aim to replace the numerous excellent texts on the respective subjects authored by eminent scholars. Instead, it aims to provide an introduction to the process of fires that have gone out of control and means of protection against fires. It further aims to bring together fundamental knowledge that a 'student' or a practitioner of fire science and fire protection or of industrial safety may find necessary. I shall be happy if they are able to use the information in their course study or profession.

At the time when the baton of fire protection in India passes from fire fighters to fire engineers, there is a need to effect a quantum change in our mindsets; to bring together the learning and knowledge that architects, scientists, engineers from diverse disciplines and fire fighters possess on the subject. There is also a need to develop texts that professionals can refer for information and practice. As a student of the subject, I continue to learn and hence, gladly welcome comments and constructive suggestions from the readers for improvement of the book.

Acknowledgements

I wish to express my gratitude to my teachers Dr. M.K. Sarkar of Chemical Engineering at IIT Delhi, Dr. R.G. Zalosh and R.L.P. Custer of the Centre for Fire Safety at Worcester Polytechnic Institute (WPI) and Dr. D. Drysdale, Adjunct Professor at WPI for encouraging me to learn the subject of Fire Safety. I thank my students from India and Oman and also the participants of many training programmes conducted by me in several countries whose questions and response have enthused me to continue learning. Also, I am thankful to the production and editorial teams at PHI Learning for their efforts. Last but not least, I wish to thank Maitreyee, my wife, and my daughters, Antara and Arpita, for their encouragement, support and patience while I practised my passion on Fire Safety and worked on this book.

Akhil Kumar Das

1
Introduction

Man is the only creature that dares to light a fire and live with it.
The reason? Because he alone has learned to put it out.

—Henry Jackson Vandyke, Jr.

1.1 FIRE IN HISTORY

Humankind recognised the awesome power of fire possibly even before it learnt to light one. It was probably this power that led to the worship of fire. Social customs in India have always had fire in the centre. Other early civilisations like the Zoroastrians, the Greeks and the Romans also worshipped fire. Many customs in different societies, even of today, hold fire in esteem and have it at the centre of different ceremonies. The Olympic Games continue to mark their beginning by lighting the Olympic Torch. On the other side, arsonists have been misusing fire right from the times of our epics and they continue with their misdeeds even today. It is probably the damage potential of fire that gave rise to the saying 'Do not play with fire'.

Once people learnt to light a fire, they also learnt to use it for the basic needs of staying warm, cooking food, making tools and keeping predators at bay. While fires were being put to such use, sometimes fires went out of control and people realised the need to protect themselves from the damaging effects of fires. Around 300 B.C., in the Roman Empire, slaves were organised to form *Corps of Vigiles* to act as fire watch at night. In the seventeenth century, the French king William II ruled that all fires in homes be covered at night and hence the term *couvre fieu,* which is the origin of the modern English word '*curfew*'. Marco Polo during his travels noted the system of appointing forms of fire watch in some oriental cities. Thus it is evident that fire safety has for long been a concern of humankind but it was the 1666 Great Fire of London (that burned from September 2 to 5) marked the start of what may be termed as *modern fire protection standards*. Following the fire, the City of London adopted a complete set of building regulations. On another front, the first fire brigade was established in Edinburgh in 1824.

The discovery of fire and the way to start one laid a major foundation for human development. Today, in a world dominated by the use of energy derived from burning fossil fuels, the combustion process in its different forms holds centre stage. Efforts are constantly on to achieve greater efficiencies in the use of all forms of energy including heat derived from burning fuels. Yet, when the same combustion process occurs where not wanted or when it goes out of control, the effects are destructive.

1.2 FIRE LOSSES

In the past several decades, scientists have developed a better understanding of the fire process and much progress has been made in the areas of fire prevention and fire protection. But, all over the world, fires continue to take a heavy toll of life and property. Despite better design of structures and improved fire protection methods, homes, hotels, airports, hospitals, schools, commercial buildings, places of entertainment, industry, museums — the list is unending — continue to experience fires resulting in irreplaceable losses. With denser populations in complex premises containing high value materials, the magnitude of actual and potential losses is also high. And, apart from unintended outbreaks of fire, arson continues to be common though it is not always easy to establish. Yet, more often than not, people imagine that a fire would not happen to them. Some even believe that fires are acts of God and cannot be prevented. Some data on fire losses in India and other parts of the world are given in Table 1.1, Table 1.2 and Table 1.3.

Table 1.1 Indian fire statistics (2014–2015)

		Fireworks	*Electrical*	*Cooking gas / stove burst*	*Riot / Agitation*	*Other causes*	*Total*
2014	Cases	151	1764	3573	14	14,875	20,377
	Injury	43	174	262	5	1405	1889
	Death	150	1779	3625	16	14,083	19,513
2015	Cases	110	2485	3390	3	12,462	18,450
	Injury	45	323	83	6	736	1193
	Death	128	2255	3400	1	11,916	17,700

Source: Accidental Deaths & Suicides in India, National Crime Records Bureau, Ministry of Home Affairs, Government of India, *Website link:* http://ncrb.gov.in/StatPublications/ADSI/ADSI2015/adsi-2015-full-report.pdf

Note: 1. Categories in ADSI Reports were modified w.e.f. 2014
2. ADSI Reports from 2016 have not been published till date.

Table 1.2 Some significant fire statistics for the period 2008–2010

	Direct loss (%GDP)	*Indirect loss (%GDP)*[@]	*Total fire deaths*	*Deaths per* 100,000 *persons*
Finland	0.17	0.011	309	2.03
Sweden	0.18	0.006	370	1.49
UK	0.13	0.008	1310	0.75
USA	0.10	0.007	9709	1.11
Japan	0.12	0.006	5609	1.51

Countries selected by the author from the original statistics are those for which data are available for the period 2008–2010 and not intended to demonstrate any degree of severity.

[@] As stated in the original report, must be regarded with serious reservations — the figures are produced on widely varying bases and some of the differences look too large for credibility.

Source: World Fire Statistics (29) 2014, The Geneva Association.

Table 1.3 Civilian fire deaths in residential buildings in the US (2013–2015)

• 83% of all civilian fire fatalities occurred in residential building fires.
• Residential building fires resulting in fatalities occurred most frequently between 11 p.m. and 7 a.m.
• November had the highest incidence of residential building fires resulting in fatalities (9%).
• Cooking (40%) was the primary cause for residential building fires that resulted in fatalities.
• 37% of civilian injuries in residential buildings resulted while attempting to escape and 31% were while sleeping.
• 90% of fatalities from residential fires involved smoke inhalation and thermal burns.
• Bedrooms (51%) were the leading location where civilian fatalities occurred in residential building fires.
• Males accounted for 58% of injuries and women accounted for 42% of the injuries.
• 37% of civilians injured in residential building fires were aged 50–69, 11% were under 10 years.

Source: Topical Fire Report Series, Volume 19, Issue 3 / December 2018, FEMA, USA.

Instances of fire are generally reported and recorded only when they turn out to be major loss events. In addition, statistics available from insurers represent only insured assets and many fires go unreported and unrecorded particularly in the Indian environment. Further, except perhaps in the case of insured assets, fires may not be thoroughly investigated to determine either their causes or the extent of loss. Even in insurance statistics, the quantum of loss can be related to the type of insurance cover that insured take. With such a range of variances, available fire statistics do not truly represent ground truth. Nevertheless, statistics can indicate trends and provide bases for decision making provided the database is reasonably wide (Table 1.1).

Losses due to fires can be divided into 'direct' losses covering damage to the premises, its contents, injuries suffered by occupants or others and fatalities (if any). At times these losses may spread to adjacent premises. In addition, there may be 'indirect' or 'consequential' losses such as lost business opportunities, idling equipment and personnel, costs for reconstruction and resumption of activities, medical costs, public liabilities, pain and trauma. Experience shows that in the case of major fire and other loss incidents, the total magnitude of indirect (also called *hidden*) losses are much larger than direct visible losses. This view of visible versus hidden losses has often been compared to an iceberg where only the tip of the iceberg is visible above the surface of water while a larger portion remains submerged (Figure 1.1). However, the concept of the iceberg of losses is not borne out by the statistics reported in Table 1.2. As the report itself points out, the figures of indirect losses must be regarded with serious reservations as the bases of computing these figures vary widely.

When estimating the impact of loss incidents, one should remember that while insurance may to a large extent recoup financial and business losses, losses like that of life, idle time of people and facilities, and the materials lost are irreplaceable.

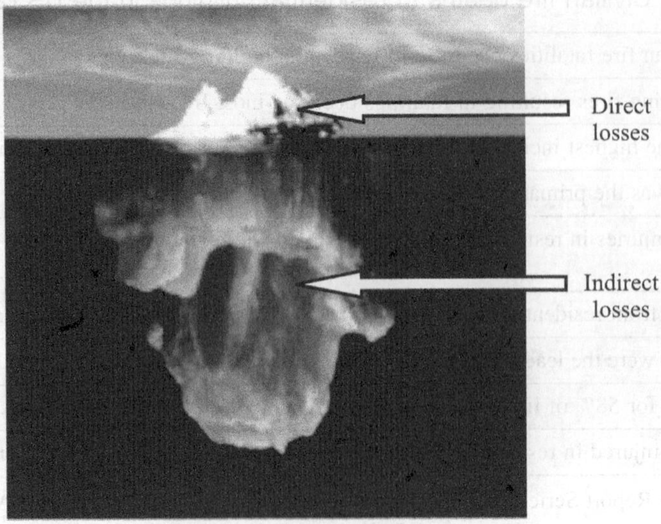

Figure 1.1 Iceberg of fire losses.

1.3 FIRE CAUSES

As will be discussed in later chapters, for a fire to start, three elements, viz., fuel, energy (commonly, heat) and a combustion supporting medium (commonly, air) are essential. Generally, 'causes' of fire are attributed to sources of heat. Thus, fire causes have been listed as

- Smoking
- Electrical appliances
- Electrical arcs and sparks
- Static charges
- Welding and cutting
- Cooking appliances (all types)
- Fireworks
- Friction
- Mechanical sparks
- Runaway chemical reaction
- Overheating
- Spontaneous
- Incendiary or malicious

Statistics compiled by different agencies have tried to assign frequencies to these or other causes of fire but their veracity cannot be confirmed due to the varying thoroughness of investigation or the competence of individual investigators.

1.4 'INTERESTED PARTIES' IN FIRE SAFETY

Fire safety should surely be everyone's concern and this includes the general public in the interest of their own safety, but there are some categories of people who have a greater responsibility for fire safety. These include law makers and enforcers, standard making bodies, fire brigades, insurance companies, institutes designing and delivering related courses, fire protection equipment manufacturers, engineers from different disciplines, fire research scientists, etc. The general public perception of fire safety in India, however, is that it is the responsibility of the fire brigades alone.

1.4.1 Role of the Fire Brigade

Ownership of the public fire services in India varies widely and so do their administration, resources and capabilities. In most Indian states, the public fire services are a wing of the police department and in some cities, they are under the administrative control of municipal bodies. Their role has largely been to extinguish major fires and conduct related rescue services although in some major cities they are also involved in approving building location, structures and fire protection hardware for conformity with local bylaws. They are called upon to respond to emergencies in a wide range of occupancies ranging from slums to sophisticated industries. These silent organisations are seen and heard only when their clanging bells and hooters are heard as they rush to an emergency scene. Interaction with the general public takes place only during the Annual Fire Services Week. Inadequacies rather than the effective work of these brigades are debated post-fire scenarios.

The Indian public fire services can take a cue from their counterparts in other parts of the world and propagate the message of fire prevention and preparedness. This can help enhance their professional image, bring them closer to their customers and facilitate implementation of regulations. The creation of a cadre of volunteer fire fighters especially in the semi-urban and rural areas also merits consideration.

Industrial fire brigades in the high-risk sector like hydrocarbon processing, power generation or aviation, on the other hand, are well equipped to manage potential emergencies in their respective premises and are sometimes called in to assist civil authorities as well.

1.5 FIRE PROTECTION STANDARDS AND PRACTICE

Recognising the need to set up norms for fire safety, Governmental and professional agencies across the world have endeavoured to establish standards for safety of both people and assets. In India, the Bureau of Indian Standards (BIS) addresses fire safety through the National Building Code. It also has a number of standards (both specifications and codes of practice) on fire safety, a list of which are given in Appendix 3. Some states and local bodies also have fire regulations and by-laws. But it is not always that these standards or regulations are addressed in engineering and architectural plans. The problem is acute in urban areas with pressure on land and resultant congestion leading to a disregard for fire safety by both builders and occupants to the extent that even access for emergency vehicles may not be available.

The National Fire Protection Association (NFPA) of the United States has been in the forefront of fire protection from its inception in 1896. Through groups of specialists from different areas, NFPA has developed standards (National Fire Codes or NFC) for fire protection and related subjects. These comprehensive standards are used or adapted in many countries of the world. NFPA has also worked to increase public awareness for fire prevention.

Recognising the hazardous nature of its activities, the petroleum industry worldwide has developed its own norms for safe operations. In this respect, American Petroleum Institute (API) has some of the earliest industry-wide fire and safety standards. Institute of Petroleum (IP), UK also has its own codes but they are not as comprehensive as those of API. In the Indian context, to their credit, the high hazard sectors of petroleum, aviation and nuclear power are probably among the best regulated in terms of fire safety as their fire protection norms and practices sometimes go beyond statutory provisions. The Oil Industry Safety Directorate (OISD) has focused its attention to different areas of safety including fire safety in the hydrocarbon industry and developed standards that are applicable to both upstream and downstream sectors. These are in addition to the provisions of statutes like the Petroleum Act (1932) and the Petroleum Rules (1978), the Explosives Act and Rules, the Cinematography Act and Rules, the Factories Act (1987) and Rules (1987), etc.

Agencies like the International Civil Aviation Organisation (ICAO), International Maritime Organisation (IMO) and the International Atomic Energy Agency (IAEA) also have guidelines on fire safety provisions and emergency management for facilities in their specific areas.

1.6 HOW MUCH SAFETY IS ENOUGH?

People often ask, 'How much safety is safe enough?' The response to that question can be another question, 'How safe do you want to feel?' In fact, sufficiency of safety from fire is a subjective judgement depending on numerous factors including personal and societal perceptions. While the statutory and the obvious minimum fire safety must be provided in all facilities, for maximum, the sky is the limit.

As a first step, one has to identify all potential hazards of an activity and assess their risks. Appropriate prevention measures start with selecting the technology and a location and layout for the activity. Thereafter it is the realm of protection. For instance, modes of segregating hazards as well as protecting them against fire are different for a flammable liquid storage and an electronic data processing facility.

After identifying hazards, one should recognise the loss potential from these hazards and resources available to handle them such that these resources can be effectively utilised to reduce potential risk, install protection measures and establish preparedness plans. These are aspects falling under the purview of risk management and are discussed later in this book.

1.7 CONCLUSION

As has been said, 'Fire is a good servant but a bad master'. In most cases (except perhaps in instances like lightning strikes), fires are not simply acts of God but are caused by human activity. There is thus a need to understand fire causes and to lower fire risks by reducing both their likelihood and their severity through design, by the use of protection hardware, through risk transfer and by being prepared to manage fire and related emergencies like explosions. In fact, fire safety essentially is a multi-disciplinary subject. Different aspects of the fire process and the fundamental aspects of fire protection will be discussed in the following chapters of this text.

QUESTIONS FOR REVISION

1. What do you understand by the terms *direct* and *indirect* fire losses? Explain the terms using the iceberg analogy.
2. Explain the term *cause of fire* as commonly understood. Prepare a list of fire causes and use examples to explain and differentiate them.
3. What are the sources of fire statistics in India? Using data from different sources analyse the agreement or lack thereof in fire statistics.
4. Discuss the role of public fire brigades in India and the scope of enhancing their role.

2

The Fire Process

The most tangible of all mysteries — fire.

—Leigh Hunt

2.1 INTRODUCTION

The word *fire* brings to mind flames leaping out from burning materials and of hot, glowing surroundings. Yet, not all fires may be flaming and, even in the case of flaming fires, the character of the flames could be different. At one extreme, we may have very slow surface burning known as *smouldering*. At the other end, if the state of the fuel, energy of the ignition source and prevailing conditions permit rapid burning, an *explosion* may ensue. The commonality in all these processes is burning, which, in the language of science, is known as *the process of combustion* and the initiation of which is known as *ignition*.

Processes like rusting or digestion also involves reaction with oxygen. However, in rusting and digestion, oxidation occurs at very slow rates as compared to combustion and the rate of heat generation is low. Fires, on the other hand, are fast reactions with release of appreciable quantities of heat energy. Combustion is an oxidation reaction but not all oxidation reactions are combustion reactions.

Combustion involves the reaction of a fuel commonly with oxygen and the release of heat. This can be represented by the equation:

$$\text{Fuel} + \text{Oxygen} \xrightarrow[\text{(Activation energy)}]{\text{In the presence of heat}} \begin{matrix}\text{Products of}\\ \text{combustion}\end{matrix} \quad \begin{matrix}\textit{(With release of}\\ \textit{heat of combustion)}\end{matrix}$$

The chemical equation below illustrates the combustion of carbon:

$$C + O_2 \xrightarrow[\text{(Activation energy)}]{\text{In the presence of heat}} \begin{matrix}CO_2 + \text{Heat of}\\ \text{combustion}\end{matrix}$$

There are exceptions to this common phenomenon such as when magnesium may burn in an atmosphere of nitrogen to give out magnesium nitride:

$$4Mg + 3N_2 \xrightarrow[\text{(Activation energy)}]{\text{In the presence of heat}} 2Mg_2N_3 \quad \begin{matrix}\textit{(With release of}\\ \textit{heat of combustion)}\end{matrix}$$

2.2 FIRE TRIANGLE

Simply put, for a fire to start, three items are essential, viz., fuel in its right form and concentration, adequate energy (normally in the form of heat) and a combustion-supporting medium (normally air). Traditionally these three items have been represented as the sides of the ***fire triangle*** as shown in Figure 2.1.

Figure 2.1 Fire triangle.

2.3 FIRE TETRAHEDRON

While the fire triangle can be used to explain the ignition process to an extent, it cannot adequately depict fire growth or extinguishment. The fire process involves the formation of unstable intermediates in a chain of reactions which occur in the flame region. The dark area between the fuel, e.g., a piece of paper or a candle wick and the visible flame is the region where the vapour formation takes place. In the case of a gas, this state already exists. As long as the flame chain reaction persists, the fire is sustained. This led to the concept of the fourth element of the fire process, viz., *flame chain reaction* and *fire tetrahedron* (a tetrahedron is a three-dimensional figure represented by a pyramid with a triangular base BCD and three triangular sloping faces, viz., ACD, ADB and ABC) (Figure 2.2).

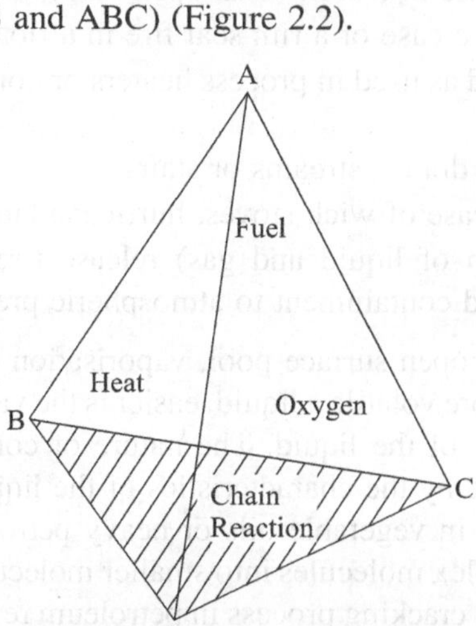

Figure 2.2 Fire tetrahedron.

2.4 FUELS

Fuels may exist in all three states of matter and their ignition sequences differ. For ignition of solid or liquid (condensed phase) fuels, the following conditions are required:

- Adequate quantities of vapour or gas have to be generated by applying energy which may be thermal alone or a combination of mechanical (e.g., spray) and thermal.
- Mixing of these vapours or gases with an oxidant like air should be in a combustible ratio.
- Adequate energy for ignition.

2.4.1 Gaseous Fuels

Gaseous fuels do not need to go through the process of vaporisation unless they are held in a liquefied state, e.g., liquefied petroleum gas (LPG) or liquefied natural gas (LNG). When the release is in the form of a turbulent jet, air will be drawn into the stream along the length of the jet. In case of a slow release, mixing with air is primarily by diffusion. Once a jet release is ignited, the point of release and the source object as well as target objects can be affected by thermal radiation and may further lead to a *boiling liquid expanding vapour explosion* (BLEVE), a scenario that will be discussed later.

When a liquefied gas is released to atmospheric pressure, it draws its latent heat of vaporisation both from the released materials and from the surrounding structures and atmosphere. This may cause frosting on solid surfaces and on the released liquid pool (if there is pool formation) as also condensation of atmospheric moisture to produce a visible mist very similar to the mist seen on winter mornings. A large release of liquefied gas can cause a flammable vapour cloud to form with the potential of a vapour cloud explosion (VCE).

2.4.2 Liquid Fuels

Liquid fuels can burn in different modes and produce different results. These include:

- Pool burning as in the case of a spill or an open surface tank.
- Annular ring fire as in the case of a rim seal fire in a floating roof tank.
- Sprays or atomised liquid as used in process heaters or boilers or release from pressurised storage.
- Flowing fires as in open drains, streams or stairs.
- Wick burning as in the case of wick stoves, hurricane lanterns or candles.
- Two-phase (combination of liquid and gas) release fires and possibly jet fires when released from pressurised containment to atmospheric pressure.

When heat is applied to an open surface pool, vaporisation occurs at the surface through a physical change of state. The more volatile a liquid, easier is the vaporisation. Convective currents are also set up within the body of the liquid. The nature of convective currents set up within the mass of liquid is determined by the characteristics of the liquid as well as its containment.

In some cases though, like in vegetable oils or heavy petroleum fractions, intense heating leads to breaking of larger complex molecules into smaller molecules that transform to the vapour state. This is used in the thermal cracking process in petroleum refining to obtain lighter fractions like LPG, naphtha and kerosene from heavy residues.

When atomised, a liquid fuel can burn throughout its mass to resemble a turbulent gas flame. This is the scenario when a liquid fuel is fired into process heaters, boilers or compression ignition engines.

In the case of wick burning, the fuel rises up the wick by capillary action before vaporising followed by mixing with surrounding air to form a flammable vapour–air mixture that can be ignited.

2.4.3 Solid Fuels

Solid fuels can have vastly different characteristics. When fuels like wax or sulphur are heated, they will first melt and then vaporise like liquid fuels. A vast majority of solid fuels, however, are polymers that may be naturally occurring, processed from naturally occurring materials or synthetic organic materials. Their main constituents are carbon, hydrogen and oxygen. In addition, these may have varying amounts of some other elements like nitrogen, sulphur or halogens and metals or metal compounds. These may exist in different forms ranging from bulk through sheets and laminates to shavings and particulates. Polymers may be classified as straight chain, branched chain, cross-linked or cyclic types. The effect of heat on each type is different. In some cases, the chain breaks to release volatiles; while in some other cases, the breaking of the chain may produce reactive smaller polymers that undergo polymerisation with each other to form highly carbonaceous char. Cellulosic and thermosetting materials convert directly from solid to vapour state when heated. The process of thermal decomposition from large complex molecules to smaller ones is known as *pyrolysis* and involves destructive distillation and vaporisation with chemical change. It occurs throughout the fuel mass where heat can reach and smaller molecules vaporise. Heat transfer is by conduction and, unlike in the case of liquids, no convective currents are set up.

In the case of wood, initially there is release of moisture causing the wood to dry up. Thermoplastic materials, on the other hand, deform on heating and some may even melt before decomposing. Some solid fuels may vaporise completely leaving no residue but generally, some residue will be left either as carbonaceous char or as inorganic residue commonly known as *ash*.

Char is an insulator, retards inward heat travel and reduces decomposition from inside the matrix, though combustion on the char surface itself continues. In fact, some flame retardant additives induce charring and thus make materials flame retardant.

In summary, the paths by which condensed fuels may release vapours for ignition are shown in Table 2.1.

Table 2.1 Some modes of vaporisation of condensed fuels

Original state	*Route to vapour state*	*Example*
Solid	Sublimation	Camphor
	Pyrolysis and evaporation	Wood and thermoset plastics
	Melting and evaporation	Wax and sulphur
	Melting, pyrolysis and evaporation	Polyurethane foam
Liquid	Evaporation	Petrol and methanol
	Pyrolysis and evaporation	Vegetable oils and heavy petroleum oils

2.5 SMOULDERING

A sheet of paper or a piece of wood placed close to a burning incandescent lamp for extended time can cause the material to discolour and vapours are evolved. This can be seen in other naturally occurring organic materials also such as cotton or jute fabrics and sawdust. These are all porous combustible solids and they burn through slow char formation, known as *smouldering*. Substances that are prone to melt and shrink from the heat source, e.g., thermoplastics like polyethylene or polyester fibres do not smoulder.

Once smouldering starts, it continues till char starts to form and then glow. In due course, smouldering may spread on the surface and also advance inwards into the bulk of the material. The chain of events, viz., discolouration, evolution of vapours, glowing char formation and production of residue continue to take place till conditions are right.

Smouldering is seen to occur only in solid fuels. However, there is a special case of smouldering involving liquids observed in industry. Heavy petroleum oils that may enter into porous insulation installed on hot surfaces can smoulder in situ leading to hard char formation (coking). Lighter hydrocarbons entering insulation tend to evaporate and disperse.

2.6 METAL FIRES

Deformed metal structural members or products are often seen in an aftermath of fire. Such deformations generally occur due to sustained high temperatures encountered in a fire. Metals generally are not fuels, though many metal salts are known combustibles. Further, strange though it may seem, some metals can also be fuels when the conditions are 'right'. Sodium catches fire if exposed to air. Powders of aluminium and magnesium can also burn vigorously and are widely used in explosives and pyrotechnics. Magnesium has a special property that it can burn in an atmosphere of nitrogen. Thus, some metals can also be fuels but the process of combustion involving them cannot be generalised and procedures of fighting metal fires also vary.

2.7 COMBUSTION OF DUSTS

When combustible dusts are present in the form of a bed, they can burn as a fuel bed as in the case of saw dust or grain husk. Flames on a burning pile of saw dust or grain husk develop a smouldering front that progresses inwards. On the other hand, many combustible dusts are prone to explode, if dispersed as a cloud (See Appendix 1, A1.3). For dust explosions to occur, the concentration of dust in air should be within 'explosive limits', which is similar to the flammable range of vapours or gases and be accompanied by adequate source of energy such as a spark or spontaneous heating. Dust explosions are also known to occur in series where a primary dust explosion can dislodge dust layers deposited on different surfaces and which can form another dust cloud with explosion potential (secondary explosion).

2.8 EASE OF IGNITION

As discussed above, fuels need to vaporise and combine with air before it can be ignited. This requires energy which may be thermal or mechanical.

It is well known that a log of wood is difficult to ignite and hence has to be chopped into fire wood. Consider a block of wood and shavings or sawdust made from an identical block of wood. It is established that shavings or sawdust ignite more easily than a block of wood. The ease of vapour generation depends on the ***surface area to volume ratio*** expressed as A/V where,

A is the total surface area of the fuel particles.
V is the volume of fuel.

Greater the A/V ratio, lower is the energy required for vaporisation as a larger number of particles and a greater fuel surface area have access to the heat source.

$$\text{Ease of ignition} \propto (\text{Surface Area/Volume})$$

Heavy oils can be discharged through a spray nozzle to facilitate burning. The increased ease of burning brought about by mechanically increasing surface area to volume ratio is effectively used in oil-fired or pulverised coal-fired industrial furnaces or boilers.

2.9 HEAT

As discussed earlier, fuel vapour and air need energy (generally, heat) to react and ignite. When a lighted match is applied to a candle wick, the heat from the match goes to (Figure 2.3):

(i) Soften and melt the wax which then goes up the wick by capillary action.
(ii) Generate wax vapours.
(iii) Set up currents in which the vapours combine with the surrounding air to form a flammable vapour–air mixture.
(iv) Ignite the flammable vapour–air mixture.

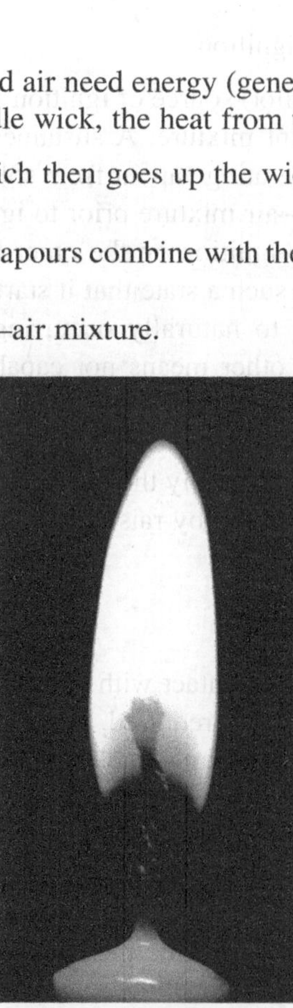

Figure 2.3 Candle flame.

All this happens in such a rapid sequence that the four (and other intermediate) steps are not individually perceptible and they all appear to take place in a flash. Once the flame is established on the wick, the match can be withdrawn as the same steps now take place with heat from the flame and burning is sustained.

Fuels other than a candle may be ignited in different ways but, once ignited, thermal feedback is essential for sustained burning. As long as there is positive thermal feedback or net heat gain (meaning that the burning process generates more heat than is lost to the surroundings) and fuel as well as adequate air are available, burning will continue. If, on the other hand, heat lost to the surroundings is high and thermal feedback is negative (net heat loss), the fire will decay or die out even if fuel and oxygen (or air) are still available.

2.10 IGNITION PROCESSES

Ignition may take place in two ways, viz.,

(i) Piloted ignition and
(ii) Spontaneous ignition or auto-ignition

In piloted ignition, an external (pilot) source of ignition like a flame or an electric spark may ignite the combustible vapour–oxidant mixture. A sustained source like a flame may perform even the preliminary functions of generating vapour from solid or liquid fuels, inducing diffusion with air to form a flammable vapour–air mixture prior to igniting it.

For spontaneous ignition or auto-ignition on the other hand, there is no external source of ignition and the material is heated to such a state that it starts to burn. This state can be reached either through internal heating due to naturally occurring or chemically induced oxidation processes or by heating it by some other means not capable of generating vapours and then igniting them. Natural fibres like cotton soaked with linseed oil and freshly mined coal are prone to self-heating and consequent ignition. On the other hand, an incandescent electric lamp or solar light passing through a convex lens can supply the energy to heat a piece of paper adjacent to it or wood to its ignition temperature to burn by raising it.

2.11 THERMAL FEEDBACK

Fuel and oxygen (of the air) may be in contact with each other in many situations but for them to react, some energy (*activation energy*) is required. For combustion reactions, activation energy implies 'sufficient' heat to take the reactants (fuel and oxygen) from their stable state to an excited state and make them reactive. Once started, the process continues till fuel and oxygen in the required proportions are available and positive heat balance (more heat is evolved by the reaction than is lost to the surroundings) exists (Figure 2.4).

The principle of positive thermal feedback can be explained taking the candle as an example. Initially the match or lighter flame will provide the heat for the wax to melt, vaporise, diffuse with air and be ignited. Once ignition is done, the match can be withdrawn and the heat of the candle flame will help sustain burning.

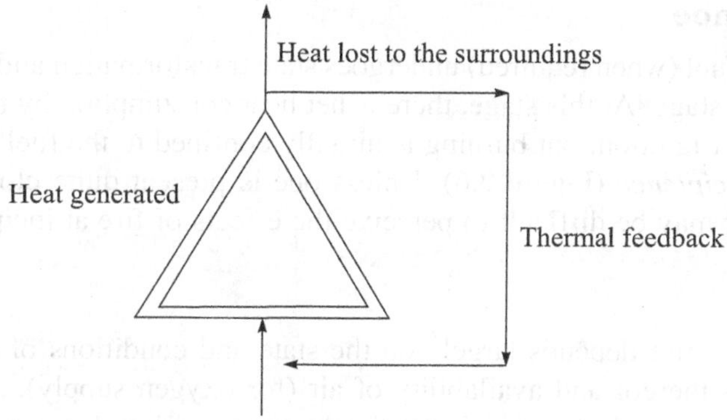

Figure 2.4 Thermal feedback.

2.12 FIRE STAGES

The life cycle of a fire starts with ignition and grows till it reaches a steady state of burning. The stages of a fire can be described in terms of heat output. From ignition and through growth there is a steady rise of temperature in the immediate surroundings, the rate of which depends on factors like the type of fuel and the availability of fuel and air. In the case of a gas fuel, heat release rate will rise sharply after ignition and the slope during growth is steep. On the other hand, for a burning piece of wood, the rate of heat release will be much slower.

After the fire stops growing, there is no further rise of temperature and steady burning with constant heat release rate is attained. For gas fuels, steady burning stage is reached almost immediately.

Thereafter, if at any stage a situation of net heat loss is reached, the fire starts to decay. This can happen if either the fuel is exhausted, or sufficient air is not available or the fire is extinguished. For a decaying gas fire, heat release rate will fall sharply while it will be slower in case of a wood fire. In the latter case though, decay can be rapid if external cooling (as from a water stream) is severe.

The three stages of a fire can be empirically illustrated as in Figure 2.5 and they are discussed a little more in the following sections.

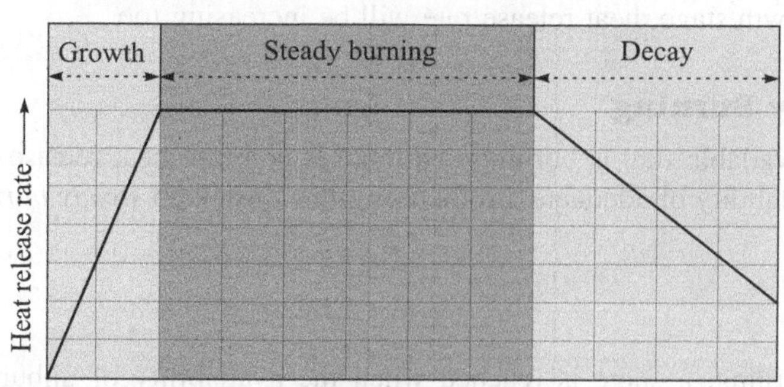

Figure 2.5 Heat release rate at fire stages.

2.12.1 Incipience

Prior to ignition, the fuel (when required) undergoes state transformation and this may be referred to as the *pre-ignition* stage. At this stage, there is net heat consumption by the fuel. Heat release commences soon after ignition, but burning is initially confined to the fuel first ignited and this stage is known as *incipience* (Figure 2.6). Unless one is present quite close to the site and in direct line of sight, it may be difficult to perceive the effects of fire at incipience.

2.12.2 Growth

The *growth* stage of a fire depends largely on the state and conditions of the fuel, its location and physical features thereof and availability of air (for oxygen supply). A premixed fuel gas and air mixture reaches steady burning immediately after ignition. For a pool of volatile liquid like petrol or methanol that gives out flammable vapours in ambient conditions, rising vapours entrain air to enter flammable range, ignite at the point where a pilot source is introduced. The flame then spreads immediately to cover the entire surface and steady burning ensues. Heavier fuels like kerosene or diesel require more heat to ignite and take longer to attain full surface burning. The larger the surface, longer is the time taken for the flames to cover the entire free surface. The resultant flame is called *diffusion flame* that can be described as a 'lazily burning flame'. The shape of the flame constantly changes and the nature of the flame (whether laminar or turbulent) will depend primarily on the dimensions of the surface. It is observed that for liquid fuel surfaces with diameter greater than 0.3 m, the flame is turbulent.

In the case of a solid fuel like a piece of wood, burning will be slower than a pool of say, petrol. Once burning is established, the flame front will progress towards the remote end. The rate of flame progression will depend on various factors such as combustibility, roughness, orientation, ambient conditions, etc.

After ignition at a point on the surface, the advancing flame front heats the surface adjacent to it and causes evolution of vapours, the ignition of which leads to further advance of the flame front and evolution of vapours. The process continues till the flames reach the other end of the surface but the rate of spread may be so rapid as to appear that the spread over the full surface was immediate (Figure 2.6).

Thereafter, growth of the fire may be from the material first ignited to other combustible materials in the adjoining space and this may be by direct flame impingement, by flying embers or by one or more modes of heat transfer.

During the growth stage, heat release rate will be increasing too.

2.12.3 Steady Burning

At this stage, all available fuel is burning and there is constant heat release rate from the fire. Depending on availability of adequate air, there is either sustained *steady burning* or decay may ensue.

2.12.4 Decay

After sustained burning, a stage is reached when the availability of unburnt fuel and hence evolution of vapours subsides (Figure 2.6). At this stage, the fire starts to *decay* and eventually dies out. Decay can also be set about through extinguishing actions.

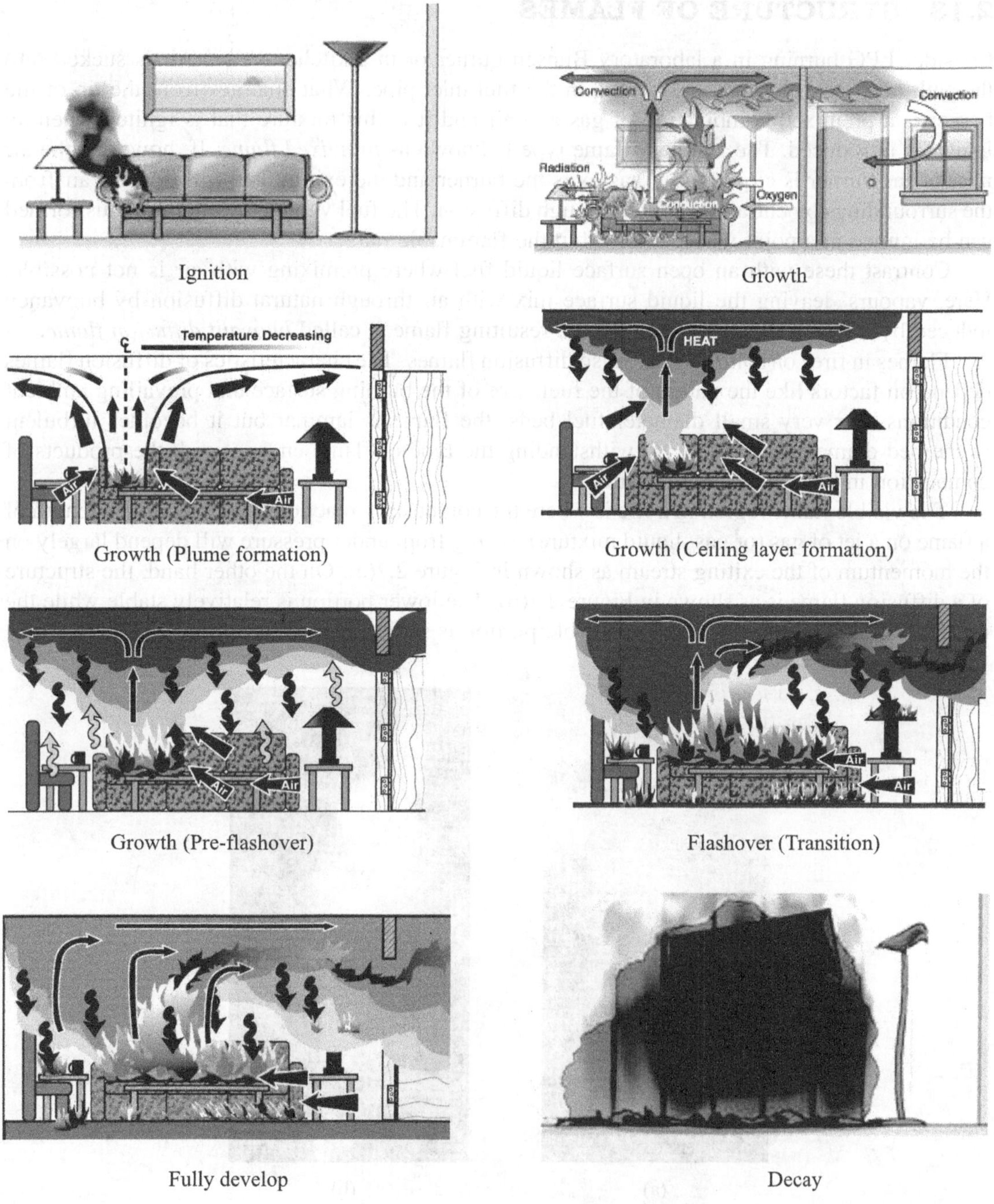

Figure 2.6 Stages of a compartment fire.

2.13 STRUCTURE OF FLAMES

Consider LPG burning in a laboratory Bunsen burner or in a kitchen stove. Air is sucked into the path of the gas through an opening in the fuel inlet pipe. What emerges from the tip of the burner is a premixed combination of gas and air and it is this mixture that is ignited when an igniter is introduced. The resultant flame type is known as *premixed flame.* If, however, the air inlet of the burner is closed, only gas exits the burner and the exiting gas will draw in air from the surroundings by entrainment and through diffusion. The fuel vapour–air mixture thus formed can be ignited at a point where it is within the flammable range.

Contrast these with an open surface liquid fuel where premixing with air is not possible. Here, vapours' leaving the liquid surface mix with air through natural diffusion by buoyancy induced by the rising fuel vapours and the resulting flame is called *buoyant diffusion flame*.

Flames in fires on solid fuels are also diffusion flames. The characteristics of diffusion flames depend on factors like the nature of the fuel, size of the burning surface and prevailing ambient conditions. For very small diameter fuel beds, the flame is laminar but it becomes turbulent as the bed diameter increases. Notwithstanding the type of flame encountered, the products of combustion in both cases are the same.

The visible flame region in a fire is where the combustion reactions take place. The length of a flame on a jet of gas (or gas–liquid mixture) exiting from under pressure will depend largely on the momentum of the exiting stream as shown in Figure 2.7(a). On the other hand, the structure of a diffusion flame is as shown in Figure 2.7(b). The lower portion is relatively stable while the upper portion is seen to flicker. The visible portion is said to have a mean flame height.

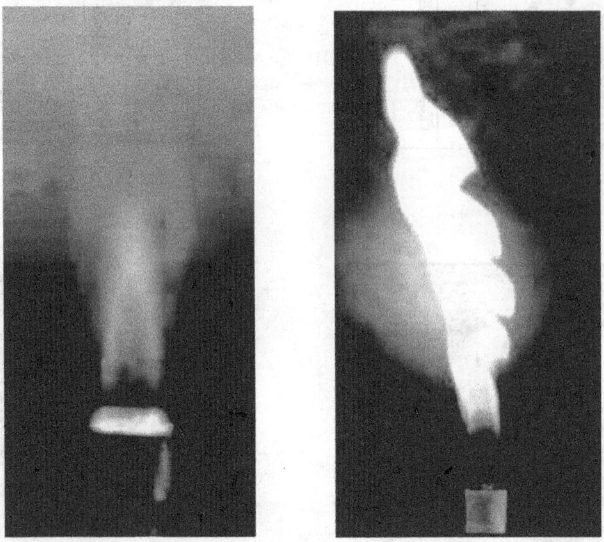

(a) (b)

Figure 2.7 Structure of different flames in case of pool fires.

Flame heights of diffusion flames in case of pool fires have been estimated to range between 1.5 and 2 times the diameter. The plumes from such fires can generate very high vertical velocities with significant air entrainment. The high velocity of a rising plume and the upward momentum of the products of combustion from large pool fires can challenge fire fighting operations. Strong crosswind conditions can both increase air entrainment and tilt the flames and render downwind facilities vulnerable.

2.14 FLAME HEIGHT/LENGTH

The shape and size of a flame largely determines the damage potential of a fire. In the case of large fires, thermal radiation is a major hazard for exposed assets and the shape and size of the flame is a major factor thereto. Hence, much research has been done to define and understand the features of a flame. Although the buoyant flame in a free-burning fire is dynamic with constantly varying features, there is general agreement that the flame can be assumed to be a stable cylinder. The upper regions of a flame are seen to be more unstable than the lower region. The base of the cylinder is assumed to be the diameter of a liquid fuel pool (in case of a pool fire) or an effective diameter for fuels present in some other configuration. Flame height (also known as *flame length*) depends on numerous parameters like the type and configuration of fuel, ambient conditions, bounding features, ventilation, etc. Further, the degree of turbulence in the flame will determine the air entrainment which too will affect the flame height. It may be mentioned that while ***flame length*** and ***flame height*** may be synonyms in the case of vertical flames, they may not be the same when the flame is tilted due to prevailing ambient conditions (Figure 2.8).

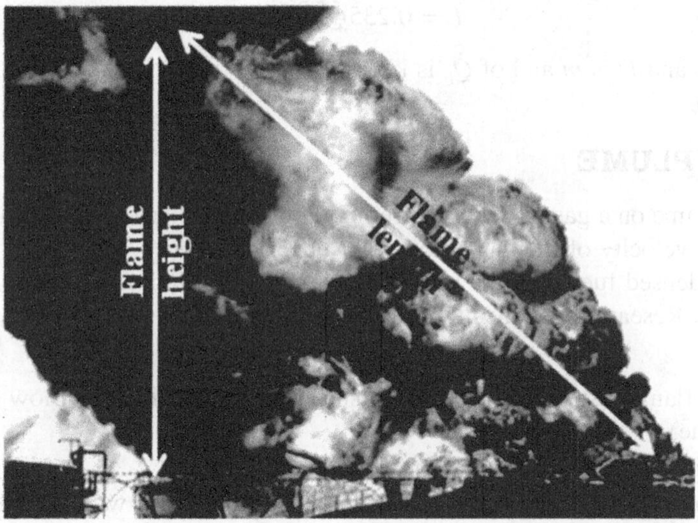

Figure 2.8 Flame length versus flame height.

Estimating the height and hence the vertical surface area of the flame cylinder is important because the thermal radiation from the flame depends on it. The exact contribution towards thermal radiation will depend on the angle of tilt of the flame cylinder, the separating distance between the flame and a target object, view factor, attenuating conditions, etc. Much work has been done in this respect particularly with liquid fuel pools and the common experimental set-up of wooden cribs and different equations have been suggested. Heskestad proposed a height to diameter ratio that is fairly independent of the fuel type or configuration:

$$\frac{L}{D} = 15.6\, N^{1/5} - 1.02 \tag{2.1}$$

where L is the mean flame height, D is the diameter of the fire (in case of circular fuel surfaces and effective diameter calculated from the equation $A = \pi D^2/4$ in other cases) and N is a dimensionless parameter defined as

$$N = \frac{\left(C_p T_\infty\right)\left[\dfrac{\dot{Q}_c^2}{D^5}\right]}{\left[\left(g\rho_\infty^2\right)\left(\dfrac{\Delta H_c}{r}\right)^3\right]} \tag{2.2}$$

where C_p is the specific heat of air, T_∞ and ρ_∞ are temperature and density respectively of the ambient air, g is the acceleration due to gravity, ΔH_c is the heat of combustion, r is the mass stoichiometric ratio of air to volatiles and $\dot{Q}_c$ is the total heat release rate given by $\dot{Q}_c = \dot{m}_f H_c$ where $\dot{m}_f$ is the mass burning rate of the fuel.

Experiments show that $\Delta H_c/r$ generally ranges around 3000 kJ/kg for most gaseous and liquid fuels. Other factors in equation (2.2) except $\dot{Q}_c$ are known at STP conditions (293 K and 760 mm Hg). Substituting known values in equation (2.1), Heskestad suggests the correlation

$$L = 0.235\dot{Q}_c^{2/5} - 1.02\, D \tag{2.3}$$

The unit of L and D is m and of $\dot{Q}_c$ is kW.

2.15 FIRE PLUME

The nature of a flame on a gas jet without premixing can be either laminar diffusion or turbulent depending on the velocity of the gas at the point of release.

Fires on condensed fuel surfaces also demonstrate both laminar and turbulent character as mentioned above. Research indicates that the ***fire plume*** in diffusion flames have three distinct regimes, viz.,

- Persistent flame just above the burner surface with accelerating flow of burning gases (flame zone).
- Intermittent flame and near-constant flow velocity.
- Buoyant plume in which velocity and temperature decrease with height.

The rising hot products of combustion of a buoyant diffusion flame set up what is termed as a *buoyant plume* that continues to rise till either temperature differentials or an overhead ceiling prevents it from rising further. As and when it reaches the ceiling, it starts spreading to form what is called a *hot ceiling layer*. If it ceases to rise due to temperature inversion, the plume starts to spread horizontally as is seen in industrial stacks. It is important to understand the phenomenon of a buoyant diffusion because much of the design of fire protection systems draws on the phenomenon.

2.16 PLUME FEATURES

At the burning surface, the cross section of the flame covers the entire surface. At a height above the fuel surface, the shape is irregular due to the flicker indicated in Figure 2.9. However, for the purpose of calculations, the flame is generally assumed to be a vertical cylinder. The plume created by the fire assumes the shape of an inverted cone that widens with elevation. It has been seen through experiments that the upward velocity is maximum along the axis of the flame and lower at the edge of the plume. This central axial velocity is maximum at a point within the flame and then falls with height. As regards temperature, it remains constant in the lower portion of the flame and starts falling with height from the intermittent section of the flame.

Behaviour of the plume depends largely on ambient conditions. In a compartment scenario, it will continue to rise till it reaches the ceiling. That too will depend on ceiling height, compartment size, openings like doors, windows or ventilation provisions, etc. In the open, plume behaviour will be affected by factors like air movements and prevailing winds, ambient temperature, inversion conditions, etc. (Figure 2.10).

Figure 2.9 Features of a fire plume. (Adapted from typical representation)

Figure 2.10 Fire plumes — Compartment and open.

2.17 FLAME PROGRESSION

The advancement of burning is another factor in the life and sustenance of a fire. In fact, the US Bureau of Mines test for estimating flammability limits is based on the travel of an applied flame across the length of the fuel–air mixture sample. On an open liquid pool, once a source of ignition is introduced at one point, the flame will spread across the entire surface. In a small pool, it may appear that the entire surface has got ignited at a single moment but the spread may be visible and evident on a pool of very large diameter. Surface flame spread on a solid fuel surface can in most cases be conveniently observed and the Steiner Tunnel test is a well-established method thereof. Progress of flame within a fuel–air vapour cloud in the open would depend on the concentration of the cloud, the size of the cloud and many other factors. The travel of flame front vis-a-vis the pressure wave in a pipe-like configuration is an entirely different phenomenon and where detonation is a likely outcome.

QUESTIONS FOR REVISION

1. What are the fundamental differences between flaming combustion and smouldering?
2. What are the different ways in which solid fuels generate vapours? Explain the process of pyrolysis in detail.
3. Describe the different stages of a typical fire starting from a given fuel. Explain the differences in the growth stage of a liquid or gaseous hydrocarbon fire as compared with a fire involving solid fuels like wood or paper.
4. Differentiate between piloted ignition and spontaneous ignition.
5. What do you understand by the terms premixed burning and buoyant diffusion flames? Discuss the difference in phenomenon and effects thereof.
6. What are the contributions required from heat inputs for ignition and sustained burning?
7. Compare and contrast 'pre-mixed' and 'diffusion' burning in case of gaseous fuels.
8. What is a fire plume? Explain the formation of a hot ceiling layer vis-a-vis a fire plume.
9. How does the velocity profile of rising gases in a plume vary from its axis to its edge?

3
Chemistry of Fire

How is it that one match can start a forest fire
but it takes a whole box of matches to start a camp fire?

—Christy Whitehead

3.1 MATTER AROUND US

All matter comprising the earth and its atmosphere are made of elements existing either by themselves or in combination with other elements. Combinations may again be separable mixtures of elements, compounds that exist as new entities with properties that are different from the elements themselves and mixtures of compounds. Water is a compound of hydrogen and oxygen with properties very different from both hydrogen and oxygen. Atmospheric air is a mechanical mixture of the elements nitrogen, oxygen and rare gases and compounds like carbon dioxide and water vapour. Arising out of natural and human activity, air may also contain oxides of nitrogen and sulphur, dusts of silica, carbon particles, etc. The properties of air are combinations of the properties of its components and the components can be physically separated.

Matter can exist in three states, viz., solid, liquid and gaseous. Depending on various conditions, the same material can exist in different states. For instance, water can be converted to steam or sulphur may be melted to be a liquid by the application of heat and liquefied petroleum gas (LPG) which is a mixture of propane and butane can be kept in liquid state by applying and holding it under pressure. The ease with which petrol converts from liquid to vapour is a property used effectively in internal combustion engines. By simply reversing the cause for change of state, the original state may be reached. Hence, such change is called *physical change.*

Combustible matter, i.e., fuels may also be solid, liquid or gas. They may be elements (e.g., carbon, hydrogen, sulphur), compounds (e.g., methane, propane) or complex mixtures (petrol, LPG, wood, paper, plastics). In general, combustible matter, whether naturally occurring or manmade, comprises carbon, hydrogen, sulphur and nitrogen in a wide variety. Some metals may also be present in them. There are also some metals like magnesium and aluminium that can by themselves, in some forms, undergo combustion reactions.

In contrast with physical change, elements and compounds or their mixtures can react chemically with each other when the conditions are right and give out products of the reaction. The products have properties that are different from any of the reactants. Chemical reactions bring about chemical changes. All such reactions involve matter and energy. Combustion (and hence fire) is recognised as a chemical reaction involving the interaction of a fuel commonly with oxygen.

3.2 AVOGADRO'S HYPOTHESIS

Although originally derived for gases, it was later observed that a gram-mole (gram-molecular weight or molecular weight expressed in grams) of any substance contains a fixed number (known as *Avogadro's number*) of its basic units (atoms or molecules), i.e., 6.023×10^{23}. *Avogadro's hypothesis* stated that this number of molecules occupy the same volume at equal pressure and temperature in case of all gases. At STP conditions (273 K and 760 mm Hg or 101.1 kPa), 1 gram-mole of a gas occupies a volume of 22.414 litres. Thus, 32 g O_2, 28 g N_2, 44 g CO_2 or 44 g propane (C_3H_8), all occupy 22.414 litres at STP. In the MKS system, the units used are kg-mole and m^3. For instance, 0.032 kg O_2 will occupy 0.022414 m^3 at STP. This fundamental relationship is useful in fire computations.

3.3 WORK VERSUS ENERGY

The first law of thermodynamics concerns the conservation of energy. Also, by the normal definition of physics, work is done when a force moves through a distance. In mathematical language, it is described (in differential form) as

$$dw = F \cdot dx \tag{3.1}$$

This can be illustrated through the movement of a piston of cross-sectional area *A* through a distance *dx* in a cylinder against a pressure *P* (Figure 3.1).

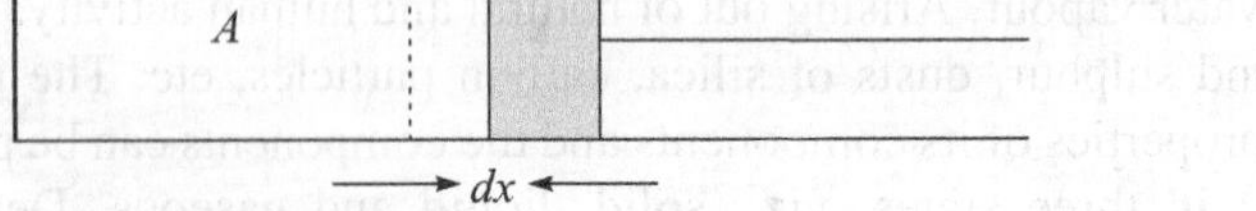

Figure 3.1 Work done by a moving piston.

Here, work done is

$$dw = P \cdot A \cdot dx \tag{3.2}$$

or

$$dw = P \cdot dV \tag{3.3}$$

where dV represents the change in volume by a cross-sectional area *A* through a distance *dx* such that $dV = A \cdot dx$

Thus, total work done can be represented as

$$W = \int_0^f P \cdot dV \tag{3.4}$$

Work is done when energy is expended. When work is done on a 'closed' system that is insulated in all respects from its surroundings, there will be a change in the energy of the system. This change of *internal energy, U*, of the system is given as

$$\Delta U = q - w \tag{3.5}$$

where, q is the heat transferred to (or from) the system and w is the work done by the system. In differential form, this relationship is expressed as

$$dU = dq - dw \tag{3.6}$$

Using equation 3.1, equation 3.6 may be written as

$$dU = dq - P \cdot dV \tag{3.7}$$

If there is no change in volume, i.e., there is no change in position of the piston in our example, $P \cdot dV = 0$ and overall, equation 3.7 can be expressed as

$$\Delta U = q_v \tag{3.8}$$

where q_v is the heat transferred at constant volume.

So far we have referred to internal energy only in terms of its change. *Internal energy* of a system represents its total energy and includes the kinetic energy due to the motion of its molecules as also the potential energy due to vibration and electric energy of atoms within the molecules and crystals that comprise the system. The amount of heat transferred due to temperature gradient between two bodies represents the change of internal energy and the work performed (equation 3.5).

Fires generally involve generation and expansion of resultant gases at constant pressure. Hence, work done in this process can be represented from equation 3.4 as

$$W = P \cdot (V_f - V_i) \tag{3.9}$$

Change in internal energy of the system can then be shown as

$$\Delta U = q_p - P \cdot (V_f - V_i) \tag{3.10}$$

or

$$U_f - U_i = q_p + PV_i - PV_f \tag{3.11}$$

where q_p is the heat transferred at constant pressure and can be written as

$$q_p = (U_f + PV_f) - (U_i + PV_i) \tag{3.12}$$

The factor $(U + PV)$ is termed *enthalpy* of the system at a given state and change in enthalpy ΔH is the heat transferred at constant pressure, q_p for an ideal system.

Equation (3.10) may also be expressed for change of enthalpy as

$$\Delta U = \Delta H - \Delta(PV) \tag{3.13}$$

where, $\Delta(PV)$ can be calculated using the ideal gas law $PV = nRT$ such that, if there is a change in the number of moles,

$$\Delta(PV) = \Delta nRT \tag{3.14}$$

3.4 ENERGY IN CHEMICAL REACTIONS

Chemical reactions involve either consumption or release of energy. Combustion reactions always result in release of energy that comprises heat and, sometimes, light. The subject of thermochemistry deals with the study of such energy relationships and is primary to the discussion of the fire process. This section discusses some energy aspects of chemical (and particularly combustion) reactions.

Most elements in their normal state when brought together with other elements (also under normal state) do not inherently react and need to be *activated* in some way for them to react. Thus fuel molecules require heat energy (***activation energy***) to be excited such that they can combine with oxygen of the air in a combustion reaction.

Electrons in an atom exist in orbits called *shells* at different distances from the nucleus. Atoms having eight electrons on the outermost shell are non-reactive as in the case of neon, an inert gas. When the outermost shell contains less than eight electrons (e.g., aluminium, carbon, oxygen and hydrogen, the atom is reactive as it tries to gain or donate electrons and become stable. Introduction of a little energy (ignition source) excites the electrons on the outer shells and aids the process of stabilisation.

Once a chemical reaction is initiated, it may either release heat to the surroundings (*exothermic reaction*) or draw heat from the surroundings (*endothermic reaction*). In general, oxidation reactions are exothermic while reduction reactions are endothermic. Thus, the burning of hydrogen (a fuel) releases heat to the surroundings (exothermic), while the dissociation of water into hydrogen and water through electrolysis is an endothermic reaction.

Heat of formation (ΔH_f) is defined as the *enthalpy change* when a compound is formed from its constituent elements, all reactants and products being at, and the reaction occurring at standard state (1 atm and 298 K). It determines the difference in energy content of the reactants and the final products formed by the reaction. By convention, negative enthalpy indicates that the energy content of the products is lower than that of the reactants.

$$H_2 + \frac{1}{2}O_2 \longrightarrow H_2O\ (l)$$

Heat of formation, ΔH_f (H_2O, l), i.e., for the final product (water) at 1 atm and 298 K, $\Delta H_f = -285$ kJ/mole. This includes the latent heat of vaporisation (44 kJ/mole), which is given out by the formed steam (1 mole) condensing into water. If the water formed by the reaction were to remain in the vapour state, heat of formation,

$$\Delta H_f\ (H_2O,\ g) \text{ would be} \longrightarrow 241 \text{ kJ/mole}$$

It may be mentioned that positive ΔH_f indicates an endothermic reaction implying that the final product is less stable than the reactants and has higher chemical reactivity. Acetylene has a $\Delta H_f = +227$ kJ/mole and is known to be an unstable compound. This property can be helpful in assessing risk of a chemical process.

Heat of reaction, ΔH (also known as ***enthalpy*** of the reaction) is the difference between ΔH_f of products and ΔH_f of reactants. This is Hess' law and affirms the first law of thermodynamics and conservation of energy. By definition, heats of formation of elements are set at zero at all temperatures.

Heat of combustion (ΔH_c) is the total amount of heat released when unit quantity of a fuel at standard state is oxidised completely in stoichiometric proportions to yield stable end products.

When hydrogen and oxygen react to form water, the reactants are elements and hence,

$$\Delta H = \Delta H_f\ (\text{water}) = \Delta H_c\ (\text{hydrogen})$$

As discussed for ΔH_f, the value of ΔH_c will depend on whether the final water produced is in the vapour state (net heat of combustion) or in the liquid state (gross heat of combustion) the difference between the two is the latent heat of vaporisation.

In contrast with other heats of reaction, heats of combustion can be measured by the combustion of a known amount of fuel in pure oxygen inside a bomb calorimeter. The calorimeter is designed so as to minimise heat loss and the heat release is calculated on the basis of temperature rise.

$$q = \text{mass} \times \text{specific heat} \times \text{rise in temperature}$$

Specific heat or ***heat capacity*** (C) of a substance is defined as the heat required to raise the temperature of unit mass of the substance through 1 deg K (or C). Unlike solids or liquids, however, gases have different specific heats for constant pressure (C_p) and constant volume (C_v) conditions. Thus, from equations (3.12) and (3.13) respectively, we have two heat equations for gases, viz.,

$$dq_p = dH = C_p \cdot dt \tag{3.15}$$

and

$$dq_v = dU = C_v \cdot dt \tag{3.16}$$

For an ideal gas, $C_p = C_v + R$, where R is the universal gas constant 8.31 J/mole·K. Accordingly, the unit for C_p and C_v is also J/mole·K.

The bomb calorimeter test is conducted at constant volume and, therefore, measures change in internal energy; fires generally follow constant pressure principles. Hence, appropriate corrections must be made when experimental C_v values are applied in fire calculations. It should also be noted that C_p values of gases depend on their temperature and this is important for fire calculations at higher temperatures.

3.5 OXIDATION AND REDUCTION

Although there are many different types of chemical reactions, the fire process involves *oxidation* reactions. In simple terms, it may be said that oxidation is the chemical reaction whereby oxygen is added to a substance and the nature of the substance itself changes. Oxidation occurs in many common processes like rusting where iron combines with oxygen in the presence of moisture and digestion of food where oxygen of inhaled air reacts with foods like fats and carbohydrates to supply the energy that the body needs.

$$4Fe + 3O_2 \longrightarrow 2Fe_2O_3$$

$$\text{Carbohydrates} + O_2 \longrightarrow CO_2 + H_2O + \text{Heat}$$

Fire is also a chemical reaction in which the elements present in a fuel commonly react with oxygen of the air to yield products of combustion.

$$C + O_2 \longrightarrow CO_2$$

$$H_2 + \tfrac{1}{2}O_2 \longrightarrow H_2O$$

$$C_3H_8 + 5O_2 \longrightarrow 3CO_2 + 4H_2O$$

Oxygen for combustion can also be available in the fuel itself in some cases as in the case of nitrates and chlorates of metals like potassium. In such cases, we see enhanced rates of combustion, e.g., fireworks and explosives.

In combustion reactions and hence in fires, the fuel is oxidised while the oxidising substance (or oxidiser) is reduced. It can also be said that, in combustion reactions, an oxidising agent and a reducing agent react together to produce heat energy and products of combustion that are less reactive than the reactants themselves.

$$CH_4 + 2O_2 \longrightarrow 2H_2O + CO_2 + \text{Heat}$$

The term *oxidation* is not limited to chemical reactions with oxygen as it applies to oxidation state and transfer of electrons. Oxidation involves loss of electrons (OIL) while reduction involves gain of electrons (RIG). The halogens fluorine, chlorine and bromine can oxidise hydrogen as in the following reaction:

$$H_2 + Cl_2 \longrightarrow 2\,HCl$$

3.6 CHEMICAL CHAIN REACTION IN FIRE

A chemical equation represents the overall stoichiometry of a complete reaction but does not indicate the inherent complexity of the reaction. Chemical reactions do not take place in a single step but proceed in transient stages, which sum up as the overall reaction. Many changes take place at different stages in which unstable intermediates are formed en route to completion. This is also true of combustion reactions in which the stages are said to occur in the visible flame region in a chain. The continuance of this chain of reactions is essential for sustained burning. Conversely, if the chain reaction is interrupted at any stage, the fire can be terminated as explained in the fire tetrahedron principle.

As a reaction is initiated through the addition of activation energy, the reactant molecules generate sub-molecular particles like atoms and unstable sub-atomic, reactive *ions* and *free radicals.* These particles react with each other and form other reactive particles. Thus there is a chain of formation and consumption of these particles in the transient stages of the chemical reaction that help to sustain the process till completion of reaction. In flaming combustion, the many steps of the complete reaction take place in the flame region. The different steps may be part of any of the following stages:

(i) Initiation
(ii) Propagation
(iii) Branching
(iv) Termination

In general, except for initiation, there may be multiple parallel or series of each of the other steps.

The foregoing discussion of the steps within a stoichiometric combustion reaction introduces a new dimension, viz., *flame chain reaction* and *fire tetrahedron* discussed in the previous chapter.

3.7 REDUCING ATMOSPHERE

If combustion occurs in a compartment with limited ventilation (*ventilation controlled burning*), the oxidation process is not sustained and products of incomplete combustion like carbon monoxide may be formed. These may undergo reaction to completion if air becomes available. This scenario is possible if air is allowed to enter the compartment for instance by breaking a window pane when the inrush of air leads to the phenomenon known as *backdraft* or a sudden transformation into a major fire.

3.8 VAPOUR PRESSURE OF LIQUIDS

When a stable liquid is kept open to atmosphere, vapours escape from the liquid surface. The rate of vaporisation and the quantum of vapour generated depend on the nature of the liquid, prevailing ambient conditions and any heat that may be supplied to the liquid. The pressure exerted by the vapour molecules just above the surface is called *vapour pressure* of the liquid. The temperature at which the vapour pressure equals the pressure above the liquid surface is the *boiling point* of the liquid.

If the liquid is covered, vaporisation continues only till the space above its surface is saturated with the vapours and equilibrium is established. For a pure liquid, this is known as *saturated vapour pressure* (SVP). For mixtures of liquids, there is a term called *partial vapour pressure* that is expressed as a product of vapour pressure of the pure liquids and the respective mole fractions of the components assuming that the mixture behaves like an ideal solution (i.e., Raoult's law is valid). Thus, for each component, partial vapour pressure p_i can be written as

$$p_i = \chi_i \cdot P_i^o$$

where χ_i is the mole fraction of component i and P_i^o is the partial vapour pressure of the pure component. Hydrocarbon mixtures behave very much like ideal solutions but most other mixtures do not.

3.9 FLAMMABILITY LIMITS AND FLAMMABLE RANGE

Simple generation of vapours from a liquid surface is not sufficient for ignition by a pilot source; generation must be in 'adequate' quantity. Oxidant is also required for combustion and this normally is oxygen from surrounding air that is entrained into the fuel vapours rising from the heated liquid surface by the dual processes of diffusion and buoyancy. The 'adequate' fuel quantity is represented by a minimum concentration of fuel in the fuel–air mixture that is required for flame to propagate through it. This concentration is called *lower flammability limit* (LFL). There is also an upper extreme concentration of fuel vapour in the fuel–air mixture called *upper flammability limit* (UFL) above which the inadequacy of oxidant does not permit flame propagation through it. The range of concentrations between LFL and UFL is called *flammable range.* This is illustrated in Figure 3.2.

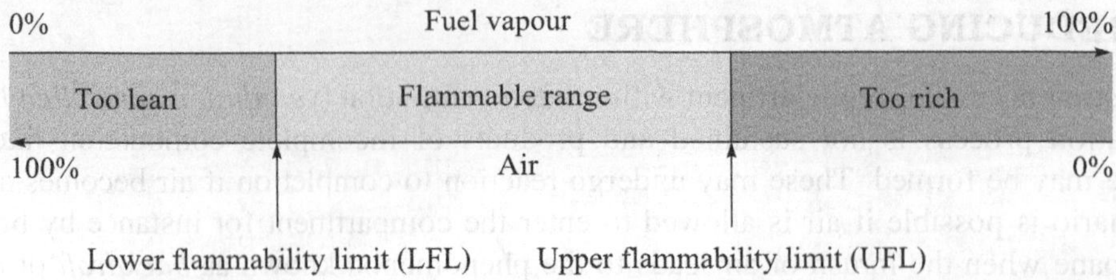

Figure 3.2 Flammable range of a fuel vapour–air mixture.

The 'flammable range' mentioned above is the concentration range in which a flame can *propagate* from the point of ignition to other sections of the mixture in the standard test apparatus. *(Note that this is in variance with general perception that flammability limits are representative of fuel–vapour air mixtures that may be ignited by a source of adequate energy).* It does not represent stoichiometry of combustion reactions as is illustrated in the simple combustion reaction of hydrogen with oxygen.

$$2H_2 + O_2 \longrightarrow 2H_2O$$

The equation shows that, for stoichiometric reaction, the required ratio of hydrogen to oxygen is 2:1. If the oxygen for the reaction is drawn from normal air (approximately 21% oxygen and 79% nitrogen), the ratio of hydrogen to air works out to 2:4.76, or a concentration of 29.59% hydrogen in air. Literature, however, tells us that hydrogen–air mixtures can support burning over a concentration range of 5–75 volume per cent. Similarly, for the combustion reaction of propane with oxygen, propane to air ratio is 1:23.8 or 4% (approximately) propane in air, while the flammable range for propane in air is 2–9 volume per cent.

$$C_3H_8 + 5O_2 \longrightarrow 3CO_2 + 4H_2O$$

Different apparatus have been developed to assess flammability limits but the US Bureau of Mines apparatus is the one that is most widely used and accepted. Flammable limits are assessed at standard temperature and pressure conditions and vary when experimental conditions are modified. In general,

- Flammable range of a fuel widens with increase in temperature.
- LFL is immune to pressure except at very low pressures.
- UFL increases with increase in pressure above atmospheric but is not sensitive to sub-atmospheric pressures.

Appendix 4 gives the fire properties of some pure flammable substances. It must be remembered that reported characteristics of fuels are derived from tests conducted under prescribed test conditions and may vary if conditions are changed. Temperature, pressure and concentration of oxidiser influence flammability limits. Higher temperatures reduce LFL and raise UFL, while higher pressures increase both LFL and UFL. Oxygen-enriched atmospheres lower LFLs and raise UFLs while oxidiser-deficient atmospheres are not flammable regardless of the fuel vapour concentration. Increasing inert gases (say, nitrogen) in a fuel–air mixture raises LFL and lowers UFL. Further, the spark energy required to ignite a flammable vapour–air mixture varies over the flammable range being minimum at stoichiometric and higher on either side of it (Figure 3.3).

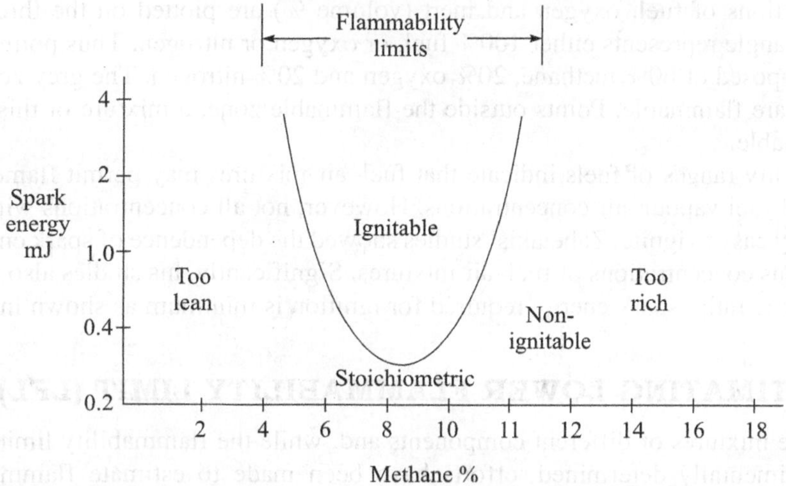

Figures 3.3 Effects of pressure and temperature on flammability of methane at atmospheric pressure and 26°C

Source: Zabetakis, 1965, US Bureau of Mines Bulletin 627.

Flammable vapour–air mixtures are actually tri-component mixtures comprising fuel vapour, oxygen and nitrogen. Accordingly, by changing the concentration of either oxygen or nitrogen, the flammability of the fuel vapour can be significantly affected. Thus, while the concept of flammable limits explains many combustion relationships, a better representation is by three-sided flammability diagrams, which is shown in Figure 3.4.

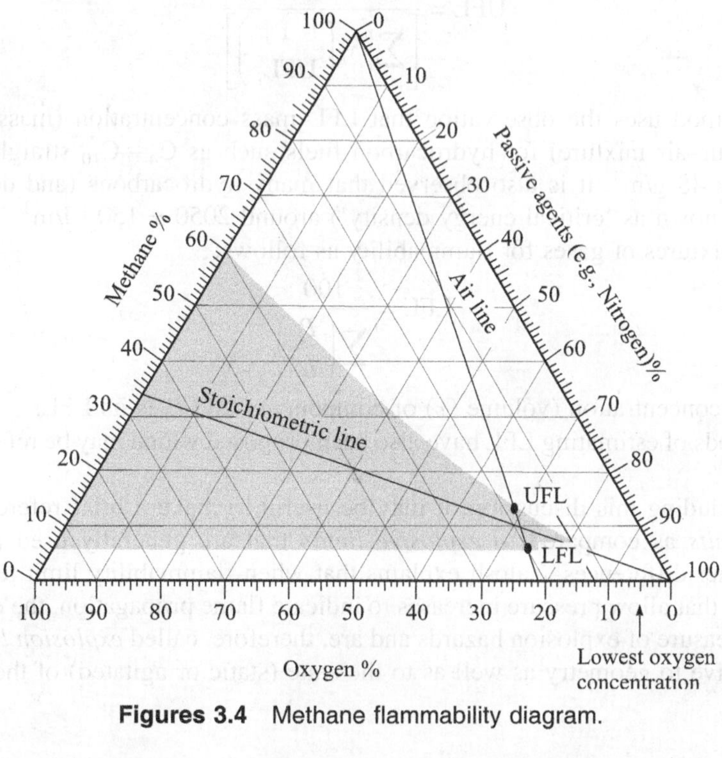

Figures 3.4 Methane flammability diagram.

Concentrations of fuel, oxygen and inert (volume %) are plotted on the three axes. Each apex of the triangle represents either 100% fuel, or oxygen or nitrogen. Thus point A represents a mixture composed of 60% methane, 20% oxygen and 20% nitrogen. The grey zone represents mixtures that are flammable. Points outside the flammable zone, a mixture of this composition are not flammable.

Flammability ranges of fuels indicate that fuel–air mixtures may permit flame propagation over a band of fuel vapour–air concentrations. However, not all concentrations within the range may be equally easy to ignite. Zabetakis' studies showed the dependence of spark energy required to ignite various concentrations of fuel–air mixtures. Significantly, his studies also indicated that at stoichiometric ratio, spark energy required for ignition is minimum as shown in (Figure 3.3).

3.10 ESTIMATING LOWER FLAMMABILITY LIMIT (LFL)

Most fuels are mixtures of different components and, while the flammability limits of mixtures can be experimentally determined, efforts have been made to estimate flammability limits based on data of components. For gas mixtures, Le Chatelier's rule has been used to estimate flammability limits on the basis that LFL (or UFL) of a fuel mixture is the reciprocal of the sum of reciprocals of the LFL (or UFL) of the components weighted by their respective mole fraction as shown below:

$$\mathrm{LFL} = \frac{1}{\left[\sum_i^n y_i \left(\frac{1}{\mathrm{LFL}_i}\right)\right]}$$

$$\mathrm{UFL} = \frac{1}{\left[\sum_i^n y_i \left(\frac{1}{\mathrm{UFL}_i}\right)\right]}$$

Another method uses the observation that LFL mass concentration (mass of fuel in unit volume of vapour–air mixture) for hydrocarbon fuels such as C_4—C_{10} straight chain alkanes is approximately 48 g/m^3. It is also observed that many hydrocarbons (and derivatives) have a ΔH_c at LFL (known as 'critical energy density') around 2050 ± 150 kJ/m^3. This permits the assessment of mixtures of gases for flammability as follows:

$$\mathrm{LFL} = \frac{100}{\sum \left(\frac{P_i}{L_i}\right)}$$

where, P_i is the concentration (volume %) of component i and L_i is its LFL.

Other methods of estimating LFL have also been proposed which may be referred in different resources.

Before concluding this discussion, it may be useful to make a brief reference to the term *flammability limits* as compared to *explosive limits* that are generally used as synonyms in spite of their finer differences. Zalosh explains that when flammability limit testing is done in pressure vessels that allow pressure increases to indicate flame propagation, the data arrived at is a more direct measure of explosion hazards and are, therefore, called *explosion limits*. Explosion limits are sensitive to geometry as well as to the state (static or agitated) of the gas under test.

Considering the potential variability in values of explosion limits, well-established flammability limits are preferentially used.

3.11 FLASH POINT

Put simply, the *flash point* of a liquid (fuel) is the lowest temperature at which it gives out sufficient vapours that can form a flammable vapour–air mixture which can flash momentarily when a source of ignition is applied under standard test conditions. Explaining a bit further, the generation of vapours at this temperature is such that once a pilot ignition source is introduced, the flame travels immediately through the vapour–air mixture consuming it and the flame goes out. Further heating will regenerate vapours which can again flash.

There are many standard methods to test for flash point of a liquid that give different results. *Closed* cup tests simulate enclosed environments as in storage tanks or closed containers and *open* cup tests simulate open conditions like a spillage. Closed cup tests represent improved precision. Flash point tests are covered as ASTM, DIN, IP, ISO and other standards. In fact, flash point values are affected by numerous factors like:

- Test method, equipment, nature and size of sample.
- Energy of the ignition source and distance from the liquid surface.
- Oxidant concentration and presence of inerts.
- Ambient pressure.
- Personal factors.

Hence, while literature may report flashpoints of various materials uniformly, ideally, the test method should be specified as the conditions are not the same for all the tests. As test method and purity of a liquid tested may vary, flashpoints should be used as guides only and not as clear demarcation between safe and unsafe. In fact, flash point has been referred as an empirical measurement rather than a property of the liquid. A flash point apparatus is shown in Figure 3.5.

Figure 3.5 Flash point apparatus.

Flash points measured at ambient pressure other than at sea level may be corrected using the following empirical equation:

$$\Delta FPt = 0.25(101.3 - P)$$

where ΔFPt is the difference in flashpoint measured at a pressure P (in kPa) and that at a pressure of 101.3 kPa (sea level).

With the aim to indicate ease of ignition, liquid fuels have been classified on the basis of their flash points. The National Fire Protection Association (NFPA) has chosen 37.8°C (100°F) as the basis: liquids with closed cup flash point lower than this temperature and maximum Reid vapour pressure (RVP) 2068 mm Hg (40 psia) at 37.8 C are called *flammable liquids* while those with flash point at or above this temperature have been termed *combustible liquids*. As it may be inferred that combustible liquids are always 'safer' than flammable liquids, it may be noted that, when handled at temperatures above their flash points, *combustible liquids* may behave as *flammable liquids*.

That is, combustible liquids can give out flammable vapour–air mixtures under prevailing ambient conditions.

In India, the Manufacture, Storage and Import of Hazardous Chemical Rules, 1989 has the following classifications for flammable chemicals:

(i) **Flammable gases:** Gases which at 20°C and at standard pressure of 101.3 KPa are

  (a) ignitable when in a mixture of 13% or less by volume with air, or;

  (b) have a flammable range with air of at least 12 percentage points regardless of the lower flammable limits.

(ii) **Extremely flammable liquids:** Chemicals which have flash point lower than or equal to 23°C and boiling point less than 35°C.

(iii) **Very highly flammable liquids:** Chemicals which have a flash point lower than or equal to 23°C and initial boiling point higher than 35°C.

(iv) **Highly flammable liquids:** Chemicals which have a flash point lower than or equal to 60°C but higher than 23°C.

(v) **Flammable liquids:** Chemicals which have a flash point higher than 60°C but lower than 90°C.

Further, the Petroleum Act, 1934 classifies petroleum products as

(i) 'Petroleum Class A' means petroleum having a flash point below 23°C.

(ii) 'Petroleum Class B' means petroleum having flash point of 23°C and above but below 65°C.

(iii) 'Petroleum Class C' means petroleum having a flash-point of 65°C and above but below 93°C.

While a liquid mass of a flammable liquid may not generate a flammable vapour–air mixture below its flash point, fine-droplet mists or aerosols of the same liquid can produce a flammable atmosphere (kerosene demonstrates such behaviour). In some extreme cases, fine-droplet mists, aerosols or foams are reported to be flammable even at temperatures less than 100°C below their flash points.

Flash points reported in literature are for pure substances and can be significantly affected by contaminants. This effect is more pronounced if the contaminant is relatively more volatile than the base liquid itself: the flash point of ethylene glycol (pure) is 111°C and just 2% acetaldehyde reduces it to 29°C.

In a similar way, adding water to a water-miscible solvent (e.g., ethanol) raises the flash point till, when sufficient water is added, it becomes non-flammable as shown in Table 3.1.

Table 3.1 Variation of flash point with concentration of ethanol in water

% Ethanol with water	*Flashpoint* (°C)
100	13.0
90	14.5
75	20.5
50	23.5
25	31.5
10	47.5
5	61.5

For solvents immiscible in water, the flash point is largely unaffected by addition of water.

3.12 FIRE POINT

In the discussion on flash point, it was observed that the pilot flame introduced consumes the flammable fuel vapour–air mixture and the flame goes out. Sustained burning (five seconds or longer) does not occur at flashpoint; it takes place at a higher temperature called *fire point* when more copious vapour is generated.

3.13 IGNITION POINT

Flash point and fire point are characteristics of liquid fuels as discussed above. There is another temperature term that is applicable to fuels of all three states, viz., solid, liquid and gas known as *auto-* or *spontaneous ignition temperature* or simply *ignition point* or *ignition temperature.* It is the temperature at which a material can self-ignite and no pilot ignition source is required.

Ease of auto-ignition depends on factors such as the chemical and physical properties of the material, size of the test sample, the test method and apparatus used, size and material of the test vessel, strength of the heat source and oxygen concentration. The *auto-ignition temperature* derived by a given test method is not necessarily representative of the minimum temperature at which a material will self-ignite in air (Figure 3.6).

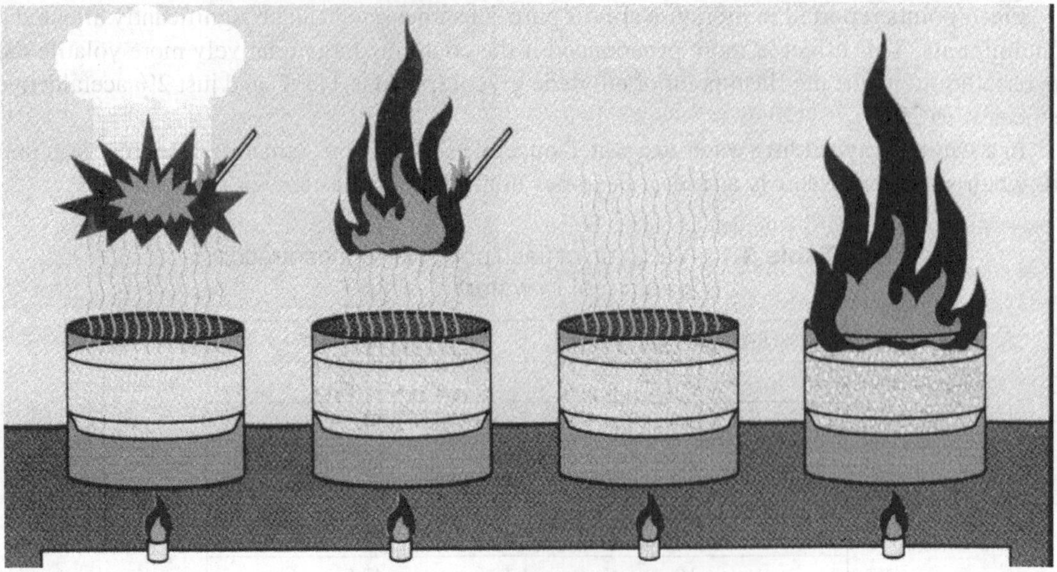

Figure 3.6 Illustrative meaning of ignition terms. (Unpublished training material)

3.14 IGNITION ENERGY

Earlier in this chapter, we have discussed the flammability limits of fuel–air mixtures *vis-a-vis* stoichiometric ratios of fuels. A point of importance noted there is that the energy required to ignite a flammable vapour–air mixture is not uniform throughout its flammability range — it is minimum at stoichiometric concentration and rises on either side.

A substance must be heated to its ignition point, or kindling temperature before it can burn. Characteristic ignition points are reported for pure substances. Although the ignition point of a substance is essentially constant, the time needed for burning to begin depends on such factors as the form of the substance, the amount of oxygen in the air and the heating source strength. A finely divided substance is more readily ignited than a massive one; e.g., sawdust ignites more rapidly than does a log. The vapours of a volatile fuel such as gasoline are more readily ignited than the fuel in bulk form.

QUESTIONS FOR REVISION

1. Using Avogadro's hypothesis, calculate the gaseous volume when 1 litre of liquid propane (specific gravity 0.5 approx.) converts to gas at STP.
2. What are the likely scenarios that may emerge in a compartment fire in case of a reducing atmosphere? What do you understand by the term 'backdraft'?

3. Discuss the variation of flammability limits of methane with variation of temperature and pressure.
4. Discuss the appropriateness of using triangular flammability diagrams in fuel–air combustion.
5. Flash point has been referred as 'an empirical measurement rather than a property of a liquid fuel'. Discuss the correctness or otherwise of this statement.
6. Differentiate between the classification of liquid fuels by NFPA and Indian stipulations.
7. 'Energy for spark ignition is the least at stoichiometric ratio.' Explain the statement *vis-a-vis* flammability limits and ignition energy.
8. What are the factors that can affect the 'auto-ignition' temperature of a substance?

4

Heat Transfer in Fires

The hardest metal yields to sufficient heat.

—M.K. Gandhi

4.1 FUNDAMENTALS OF HEAT

The process of combustion and its sustenance is all about heat transfer. Sustained fires generate heat and raise the temperature of the surroundings including any material objects and cause the fire to spread. The character of a fire and its ability to sustenance depend on the rate of heat produced, the heat conveyed to other materials and the materials present. Before going on to discuss various aspects of heat and heat transfer during fires, let us briefly review some salient features of heat as a form of energy.

In Chapter 3, 'Fire Chemistry', we have discussed the concept of work done by the mechanical movement of a piston in a cylinder. As work is to mechanical energy, heat is to thermal energy. In the same chapter, the kinetic energy component of internal energy of a body due to the vibration of its constituent atomic particles was also discussed. At an absolute temperature of 0 K (–273°C), there is supposed to be no such movement and the kinetic energy of the body is said to be zero. This is sometimes described as the *heat content* of the body and has been expressed as the integral of the specific heat of the material from 0 K to the subject temperature, or

$$q^m = \int_0^T \rho c dT \tag{4.1}$$

where,

q_m = heat content of a body of mass m

ρ = specific gravity of the material

c = heat capacity (specific heat) of the material

0, T = initial and final body temperature (K).

At 'high' temperatures (different for different materials), specific heat is said to become constant. For most practical calculations, it is reasonable to assume that specific heats of solids and liquids are not temperature dependent.

The heat content of a body has also been described as the thermal energy stored inside it while temperature is a measure of the degree of hotness. In the MKS system, the unit of heat is Joule (J) which is the energy provided by 1 Watt (W) flowing for 1 second. Expressed differently, 1 calorie (cal) is the amount of heat required to raise the temperature of 1 gram (g) of water through 1°C (from 15.5 to 16.5°C). The heat capacity of water is 1 cal/g. The heat gained (or lost) by a body is expressed as a product of its mass, its *specific heat* or *heat capacity* (a property of the material) and the rise (or fall) in temperature.

i.e., Heat gained (or lost) = mass × specific heat × rise (or fall) in temperature

or $$q = m \times c \times \Delta T \tag{4.2}$$

Thus if a copper bar weighing 10 g is heated through a temperature of 10°C, the heat gained by the bar would be

$$q = 10 \times c \times 10 \text{ J} \quad \text{(specific heat of copper = 0.385 J/g)}$$
$$= 38.5 \text{ J}$$

Another form of heat that needs to be considered is that concerned with the change of state. At its boiling point, when water changes from liquid to vapour state, there is no rise in temperature yet heat is required to bring about this change of state. The heat required for this change of state is known as the *latent* (Latin for hidden) *heat of vaporisation,* which also is the heat given out when steam converts to water. This property of water can be effectively utilised in extinguishing fires with water sprays or mists. Similarly, at the melting point of ice, ice converts to water without any rise in temperature and the heat required for melting is called the *latent heat of fusion*. Latent heat is expressed as heat per unit mass necessary to change the state of a substance and the unit for both latent heat of vaporisation and latent heat of fusion is J/g or cal/g. In the case of water, latent heat of vaporisation is approximately 2257 J/g (540 cal/g), while latent heat of fusion is approximately 334 J/g (80 cal/g).

When heat is transferred from a source to a receiver, the former loses heat while the latter gains it. In the process, the temperature of the receiver rises till equilibrium is reached and heat transfer is no longer possible. In the case of fires, heat generated by combustion is lost to the surroundings and this heat transfer will continue till either the fire is extinguished or the fuel (combustibles) is completely consumed. By convention, heat gained by a body is expressed as positive while heat lost is expressed as negative.

Fuels have heat contained within them (heat content) that is released as heat of combustion when they burn. The flow of heat is always from a source at a higher temperature to a receiver at a lower temperature and does not depend on the quantity of heat stored in a body. The analogy to this is that water flows from a higher level to a lower level irrespective of the quantity of water stored.

Like many other chemical reactions, combustion requires energy to activate the fuel components and initiate the process of burning (activation energy). As discussed earlier, heat is required to generate fuel vapours (diffusion flames), to ignite the fuel–air mixture (diffusion or premixed flames), to take the fuel to ignition temperature (surface combustion) and for sustained burning after ignition has taken place. In all cases, there is a consistent transfer of heat from one material to another through one or more modes of heat transfer viz., conduction, convection and radiation. The relative contribution by these modes depends on different factors like the fuels involved, the materials in the vicinity, the source of heat, the temperature of the source, the target receiving the thermal energy, prevailing ambient conditions, location in confinement or in the open, etc. In general, however, heat transfer during a fire is primarily by convection and radiation while heat transfer by conduction is relatively low.

4.2 HEAT TRANSFER AND HEAT FLUX

As mentioned earlier, the heat generated by combustion will necessarily be transferred to the surroundings. The rate at which this energy transfer takes place is called the *heat transfer rate* or heat flow per unit time and may be denoted as q' and has the unit of power viz., watts. Heat transfer rate from a fire depends on factors like the nature and state of the fuel, its rate of consumption, location of the fire and ambient conditions.

The term *heat flux* (q'') is used to describe the rate of heat transmitted (or received) per unit area and has the unit watts per meter squared (W/m^2). The susceptibility of people and materials to thermal conditions of a fire is largely determined by the heat flux it receives and which is known as *incident heat flux*. It is measured as the heat received per unit time per unit area of target surface in a direction perpendicular to the surface of the emitting hot surface and depends on various factors including the nature of the heat source and that of the receiver, their relative orientation (view factor), ambient conditions, the surrounding conditions and the modes of heat transfer.

4.3 MODES OF HEAT TRANSFER

The spread of fire from one body to another may materialise by:

- contiguity of a burning object with combustibles that are not burning,
- physical transport of burning objects (like sparks), or
- transfer of heat energy through one of the three modes of heat transfer, viz., conduction, convection and radiation.

The three modes of heat transfer are briefly discussed in the following sub-sections. In the case of industrial equipment such as fired process heaters or boilers, heat transfer is mainly achieved through radiation and convection. Similarly, in the case of hostile fires too, radiation and convection are mainly responsible for fire growth and spread. For elaborate explanations, readers may refer to appropriate literature.

4.3.1 Conduction

The process of transfer of heat from one body to another through physical contact is termed *conduction*. It is normally associated with heat transfer within a solid body or between solid bodies in contact with each other. A metal object heated at one end in due course becomes hot at the other end also. The particles within the body nearest the heat source get heated first and transfer heat to the cooler particles adjacent to them. The process continues till the particles at the farthest end become hot. If the first object is in contact with a second cooler body, heat is transferred by conduction to the second body as well, e.g., hand holding a metal rod being heated at one end.

The ability of a material to transfer heat by conduction is called *thermal conductivity*. This property varies widely among materials and is much higher for metals than for materials like wood. Materials with poor thermal conductivity can act as thermal insulators. Air is a very poor conductor of heat and this property is effectively utilised by providing air gaps within compartment elements such as walls.

A fire can spread by conduction through a metal wall panel, a steel door, a metal pipe passing from one compartment to another, a wooden member in contact with a hot surface, etc. Oil in a pan heats through conduction and can start burning after the oil pyrolyses and combustible vapours are released.

Heat flux (heat transferred per unit time) due to conduction from a unit surface area can be expressed mathematically as

$$q'' = \frac{-k\Delta t}{\Delta x} \tag{4.3}$$

where Δt (°C or K) is the temperature difference over a distance Δx (m) and k is the *thermal conductivity* of the material with the unit W/(m·K). In differential form, it is expressed as

$$q'' = \frac{-k\,dt}{dx} \tag{4.4}$$

This relationship is known as the Fourier's law of heat conduction. It may be mentioned that thermal conductivity of a material varies with temperature and values reported in literature are for standard temperature conditions like 0°C or 20°C.

Materials that have high thermal conductivity are generally seen to be good conductors of electricity too (Table 4.1). This is attributed to the presence of free electrons that contribute both to heat transfer as well as to the passage of electricity. In the case of insulators, conductive heat transfer is limited as it occurs only through mechanical vibrations of the particles constituting the body mass.

Table 4.1 Thermal conductivity of some building materials

Material	*Thermal conductivity* (kW/mK) *at* 25°C
Styrofoam	0.033
Wood (oak)	0.17
Gypsum board	0.17
Vermiculite	0.058
Carbon steel	43
Aluminium	205
Copper	401

It is important to note that the above discussion on conductive heat transfer is limited to steady state conduction. This represents a state of equilibrium at which the character of both the heat source and the material remain unchanged. In the case of fires, however, there is constant variation in the fire source and any conduction from such a heat source is said to be unsteady state conduction. Any analysis of such conductive heat transfer requires the use of time-dependent partial differential equations and is kept outside the scope of this text. For further details on the subject, readers may refer to sources listed in Suggested Reading appended at the end of the book.

4.3.2 Convection

When a liquid in a vessel is heated from beneath, initial heat transfer between the bottom of the containing vessel and the layer of liquid in contact with it is by conduction. Once heated, the density of the bottom layer of liquid falls and the heated lighter liquid rises while cooler, heavier liquid above comes down to establish what are called *convection currents.* Thus the rising hot liquid, through its physical movement, transfers heat to liquid at higher levels.

As in the case of liquids, gases and vapours (and their complex combinations such as smoke and mist also) can undergo heat transfer by convection. In case of a fire in a building compartment, the air in the compartment heats up and, with the hot gaseous products of combustion, establishes convection currents whereby they rise up towards the ceiling. This is the basic cause of the formation of a *hot ceiling layer* in compartment fires and can be a mode for a fire starting at grade to spread to the upper portions of a compartment. When the conditions are right, the rising hot gaseous products create a *stack effect* that tends to draw in cooler air at lower levels into the rising plume so that burning is sustained and the fire grows upwards. In the case of fires in the open, as in an open process plant, convective heat transfer from fires below can affect equipment at higher locations.

Convection necessarily involves heat transfer by movement of a fluid. However, there are two distinct types of convective heat transfer. For instance, the process of heating a liquid in a vessel as described above, wherein a hot surface makes a fluid less dense and raises it by buoyancy is called *natural convection.* If on the other hand, the subject fluid is from a continuous inflow and is absorbing heat on its path from a hot surface, the process is termed *forced convection.* In the former, upper layers will continue to get heated till the entire mass of fluid attains uniform temperature (at least theoretically). In the latter case, there is a thermal gradient as well as a velocity gradient: a layer close to the hot surface is at the same temperature as the surface and is almost stationary (*boundary layer*, velocity = 0), while there are variations both in temperature and in velocity as one goes away from the hot surface.

Fire scenarios can demonstrate a combination of both forms of convective heat transfer and analyses must address both phenomena. As convective heat transfer involves flow of fluids, and more prominently in the case of forced convection, it is good to be aware of the scope of various flow factors like Reynolds number, Nusselt number, Prandtl number and Grashof number. These are discussed with particular reference to hydraulics in a later chapter.

The transport of heat by convection raises the temperature of the surrounding atmosphere and materials and may cause a fire to spread if heat transfer is sustained. Fire spread by convection depends on many factors and one such factor is the proximity of surfaces opposite each other. If there is a wide gap between the two vertical surfaces one of which is burning, the heat released by the burning body may dissipate leaving the second surface relatively unaffected. If, on the other hand, the gap separating them is narrow, a stack effect is created thereby providing a channel for the hot gases to rise and aiding the growth of the fire. Further, the closely placed second surface may be heated adequately to start burning making the overall scenario a severe fire. Newton proposed an equation for heat transfer by convection as

$$q' = h \, \Delta t \tag{4.5}$$

where h is known as the *convective heat transfer coefficient* with the unit $W/(m^2K)$.

Unlike k in conductive heat transfer, h in convective heat transfer is not specific to any material but depends on factors like geometry and orientation of the burning materials, properties of the surrounding air, temperature difference between the source and the receiver, etc.

4.3.3 Radiation

In heat transfer by radiation, neither is the heat-emitting source in contact with the receiver (as is required in conduction) nor is there any physical movement of the surrounding medium (as in the case of convection); in fact, heat transfer by radiation can occur without any medium between the emitter and the receiver and can be described by electromagnetic theory or by quantum mechanics in terms of photons.

In radiation, the transfer of heat depends primarily on the temperature differential between two surfaces and the surface area of the hot body (emitter). It is a direct transfer of energy. Other factors that affect radiant heat transfer are the relative orientation between the emitting and receiving surfaces and the distance between the emitter and the receiver. Heat from the sun reaches the earth through radiation and hence is termed *solar radiation*. Heat transfer by radiation is a function of the fourth order of temperatures at the transmitter and the receiver and hence can have dramatic effect in very high temperature conditions. It is estimated that the temperature at the surface of the sun is around 5800 K and on reaching the earth heats the atmosphere to about 320 K.

Radiation is an important mode of heat transfer in large fires that may cause the fire to spread to other combustible materials. In the case of radiant heat transfer between two surfaces, the tendency of the cooler body to get heated depends on the type of surface of the cooler body. A polished, shining surface will reflect much of the heat energy it receives, while a rough, dark surface will tend to absorb it. This property makes white or light coloured clothing more comfortable for summer wear. Good reflective property of polished metallic surfaces is utilised in the aluminised protective suits of fire fighters. If, however there is soot deposition on the material, its reflective property reduces and the material may start absorbing incident heat.

The relative position of the surfaces is also a factor that determines heat transfer by radiation. In our discussion on convective heat transfer, we had touched on the aspect of enhanced convective currents through stack effect of two closely spaced surfaces. In closely placed parallel vertical faces (as between adjacent high-rise buildings or high-rack storage), there is also enhanced radiation and re-radiation between the surfaces.

It may lead to the spread of fire from one building to another. Similarly, in the case of large hydrocarbon pool fires, the effects of radiation may lead to growth of fire or cause damage to other assets as well.

The famous Stefan–Boltzmann equation states the energy emitted by a body as

$$E = \varepsilon \sigma T^4 \tag{4.6}$$

where σ is the Stefan–Boltzmann constant (5.67×10^{-8} W/m^{2}K) and ε is the emissivity of the surface, which is a measure of the efficiency of the surface as a radiator of heat. By convention, emissivity of a *black body* (assumed as a perfect emitter) = 1.

From the Stefan–Boltzmann equation, it is evident that radiant energy emitted is proportional to T^4 (in K).

However, not all the energy transmitted reaches the receiver surface and this depends on their mutual geometry and orientation known as their *configuration factor* (Φ). Thus,

$$q'' = \Phi \, \varepsilon \sigma T^4 \quad (4.7)$$

In addition to this, factor like *attenuation* due to conditions of the surroundings such as the presence of water vapour also play a role in the energy received. Water sprays can be applied to control burning when both the high latent heat of vaporisation of water and attenuation of thermal radiation by the water droplets are effectively utilised. On the other hand, particulates in smoke can enhance the level of thermal radiation. Further, as a fire ages, the particles agglomerate and become larger such that level of radiation is enhanced.

Some effects of incident thermal radiation on different target objects are given in Table 4.2.

Table 4.2 Effects of thermal radiation

Incident thermal radiation (kW/m^2)	*Effect*
37.5 sustained	Damage to equipment and exposed steel structures
30.0 sustained	Limit for Class 1 building materials
12.5 sustained	Melting plastic
4.0	Blistering on skin on brief exposure
1.6	Severe hot feeling
1.0	Received at sea level on the equator at noon

The net radiation heat loss rate from a hot object to its cooler surroundings can be expressed as

$$q = \varepsilon \sigma (T_h^4 - T_c^4) \, A_c$$

where,

T_h = hot body absolute temperature (K)
T_c = cold surroundings absolute temperature (K)
A_c = area of the object (m^2)

4.4 TEMPERATURES VERSUS HEAT IN FIRE

As we are very well aware by now, combustion is a rapid exothermic chemical reaction but combustion itself may be of different types. The oxy-acetylene flame commonly used for welding or cutting metals generates temperatures high enough to melt steel and produce a weld. An electric arc (as in arc welding) generates even higher temperatures. Smouldering solids like the tip of a joss stick (agarbatti) can ignite common combustibles like paper or textiles but cannot even attack the surface of a steel item. While the temperature reached in a fire is important, the total heat generated by the combustion process, as well as the heat release rate are vital parameters to determine sustenance and spread of fire.

It is also worth recalling that temperatures like flash point, fire point and ignition temperatures for various materials are tested at and reported for standard test conditions. Fires can and do take place in very different situations. We have discussed earlier that ignition is preceded by the generation of vapours from the surface of a material and its mixing with air to form a combustible mixture. This requires a minimum heat flux called the *critical heat flux* (CHF) which is specific to the material. Each material also exerts some resistance to this process which too is characteristic of the material and is known as *thermal response parameter* (TRP). The time to ignition for any material is proportional both to the CHF and the TRP of the material.

During a fire scenario, the combined flux due to the flame (both radiative and convective components) on the material and all external heat sources can cause the pyrolysis front and the flame to move forward causing the fire to spread. The rate of this spread is termed *fire propagation rate.*

The heat given out by a stoichiometric combustion reaction is the heat of combustion indicated by the enthalpy change between the heat of formation of the reactants (fuel and oxygen) and the heat of formation of the products. Fires and in fact, most combustion reactions, however, are not stoichiometric reactions and the reactions themselves may not be complete. Depending on various factors including ambient conditions, the same fuel may liberate different quantities of a range of combustion products. To address such behaviour of fires, combustion processes are described in terms of *combustion efficiency*, which is the ratio of the *actual* heat of chemical reaction to the *net heat of complete combustion.*

4.5 SEVERITY AND GROWTH OF FIRE

The severity of a fire is largely determined not as much by the total heat released till completion of combustion as by the rate of heat released (preferably referred as *chemical heat release rate*) expressed as

$$Q'c = m' \cdot \Delta H_c$$

where m' is the rate of mass loss and ΔH_c is the net heat of combustion. In real fires, however, different fuels may be burning simultaneously and also, fires normally do not reach complete combustion even if there is adequate ventilation. A part of the chemical heat release rate is transferred beyond the ignition zone through the fuel as conductive heat flux and another part as convective and radiative heat fluxes from the flame. If this heat transfer meets the criteria (CHF and TRP) referred in the foregoing section, burning and flame spread over a larger fuel surface leading to increase in flame height, chemical heat release rate and heat flux transfer and the process repeats itself for further fire growth. The rate of growth of pyrolysis is described in terms of *fire propagation rate*.

As the fire propagates and grows, pyrolysis and hence, mass loss due to gasification of the burning material also increases till it reaches a steady state. The mass loss rate depends on factors such as the heat flux reaching the burning material from the flame and external sources, the heat flux lost from the fuel surface by reradiation and the heat of gasification. The last named factor again depends on the state and nature of the fuel, ambient temperature and also material properties like melting point, evaporation temperature, specific heat and latent heats of melting and vaporisation. Tewarson provides details on a range of methods to determine the heat of gasification of different types of materials. The cone calorimeter developed by Tewarson to estimate heat release rate is a standard equipment used in FM Global and by fire researchers elsewhere.

The effects of heat released from a fire will depend on numerous factors like whether the fire is in the open or in an enclosed space, ambient conditions, nature of materials in the surroundings, etc.

4.6 SPONTANEOUS HEATING AND SPONTANEOUS COMBUSTION

In our discussion so far we have referred to preheating and ignition by an external source of ignition which provides the required energy. This phenomenon is known as *pilot ignition*.

There is another path to ignition and which is observed in moist organic materials such as freshly mined coal, hay (more likely when baled) or some dispersed vegetable oils as in oily rags. It is reported that the moisture content in the case of hay should be more than 15% for the process to occur. Spontaneous heating is the process of self-heating due to microbial growth and biochemical oxidation reactions of living organisms. As most such organisms cannot survive at temperatures above 50°C and hence may not be the only path to ignition, it is believed that some exothermic chemical reactions also take place for sustained heating and ignition to occur. Spontaneous heating requires oxygen with ventilation playing a major role — if ventilation is limited as in metal bins with lids in place, self-heating may not be sustained while plenty of ventilation will help disperse the heat generated and prevent accumulation of heat.

Self-heating of some organic materials can also occur due to processes like exothermic self-polymerisation. To develop into self-ignition, the activities must be exothermic and three factors interact to control the chain of events, viz.,

- The dimensions of the stored material,
- The ambient temperature and
- Heat conduction through the material.

4.7 HEAT RELEASE RATE

The 'size' of a fire can be a matter of individual perception. People often perceive it by the height of flames or the damage done. Fire engineers, however, have their own way of expressing the 'size' of a fire, which they do by *heat release rate* (HRR). The temperature of a single candle flame is the same as several candle flames but the HRR in the latter case is much higher than that of a single candle flame. This may be explained in terms of greater thermal feedback and thus greater sustenance and growth of the burning process. HRR is also indicative of ease of controlling a fire.

As Babrauskas puts it, HRR is not just 'one of many' variables used to describe a fire, but it is the most important variable in describing fire hazard with explosions as the only exception. The main reasons for this are:

- HRR is the engine driving a fire. This tends to occur in a positive feedback way where heat makes more heat.

- The generation of most other undesirable fire products such as smoke, toxic gases, room temperatures tend to increase with HRR.
- High HRR fires, however, are intrinsically dangerous because high HRR causes high temperatures and high heat flux conditions, which may prove lethal to occupants.

In general, a typical HRR curve has a parabolic profile with HRR peaking with time before it starts depleting. The initial steepness of the profile will depend on the type of fire (smouldering or flaming), the type and arrangement of fuel, whether enclosed or open, enclosure dimensions, ventilation, etc.

QUESTIONS FOR REVISION

1. What are the different ways in which fire can spread? Using a high rise building scenario, explain potential fire spread by the three modes of heat transfer.
2. Thermal feedback by radiation and reradiation causes the fire to grow fast. Explain the phenomenon with respect to a fire at centre, near the wall and corner of a compartment.
3. Fire growth can be influenced by proximity of two adjacent buildings – Explain. How does this compare with high rack storage in warehouses?
4. Explain the term 'incident heat flux'. How is it used to establish 'safe separation distances' between any two built vertical surfaces.
5. How is the 'size' of a fire estimated? How can HRR be used to assess the stage of a fire?

5

Hydraulics, Pumps and Primers

When tempted to fight fire with fire, remember fire fighters usually use water.

—Anon

5.1 INTRODUCTION

Hydraulics is the study of fluids. The part of hydraulics that deals with static fluids is known as hydrostatics, while fluid flow falls in the domain of hydrokinetics. As most fire protection systems are water based, it is useful to understand the fundamental theories of water flow. Although vaporising liquids and gases are also fluids and are used for fire protection, this is not included in our discussion and the emphasis in this chapter is confined to the characteristics of water in storage and in flow.

5.2 COMMONLY USED TERMS

Before going on to discuss the principles and the use of fluid flows, it may be useful to briefly discuss some fundamentals dealing with fluids.

Density (ρ) is defined as the mass per unit volume of any substance and has the unit kg/m^3 (SI units).

$$\rho = \frac{\text{Mass}\,(m)}{\text{Volume}\,(v)}$$

Often, in the case of liquids, the density of a substance is compared with that of water and expressed as a ratio called *specific gravity*. Or, expressed in another way, specific gravity is the ratio of a mass of a liquid to a mass of the same volume of water. Being a ratio, it has no units.

Specific weight (γ) is the weight per unit volume of a substance and has the unit kN/m^3 (SI units). It is known that,

$$\begin{aligned}\text{Weight} &= \text{Mass}\ (m) \times \text{Acceleration due to gravity}\ (g),\\ &= \rho v g\end{aligned}$$

Thus, for unit volume (or specific weight), we have

$$\gamma = \rho g$$

Pressure is defined as the force per unit area (SI unit kPa, kiloPascal or N/m^2).

$$P = \frac{F}{A}$$

In the case of fluids, pressure is exerted either on an object submerged in it or on the containing surfaces. For an enclosed fluid, Pascal's law states that equal pressure is exerted in all directions. In the case of an open container, downward pressure is proportional to the depth as well as the density of the contained liquid.

An important characteristic of a fluid is its ***viscosity*** (commonly termed as *thickness*), which is indicative of the resistance of a fluid against deformation due to shear stress as also of resistance to pour or flow. Viscosity is commonly assessed by the resistance to the descent of an object through a fluid by gravity. Thus vegetable oils are seen as thicker (or more viscous) than water. Similarly, lubricating oils are thicker than petrol or kerosene.

Newton postulated that for linear, parallel and uniform flow, shear stress (τ) between fluid layers is proportional to the velocity gradient ($\delta u/\delta y$) perpendicular to the direction of flow and may be expressed as

$$\tau = \mu\left(\frac{\delta u}{\delta y}\right)$$

where the constant of proportionality μ is known as the *coefficient of viscosity*. This relationship is true for all Newtonian fluids including water. The SI unit of viscosity is $kg.m^{-1}s^{-1}$ but the cgs unit of centipoise (cP) is commonly used as the coefficient of viscosity of water at 20°C is close to 1 cP. Viscosity is a temperature-dependent property and generally, decreases with rise in temperature for liquids, while for gases, it increases with temperature. The variation of viscosity with temperature is called *viscosity index*. Water used for fire protection purposes is at ambient temperature and hence, the standard value of viscosity may be used for all calculations.

The concern often, however, is more with the ratio of viscous force to inertial force (the latter is a density-dependent property of fluids) than with viscosity itself and, for this, we have another viscosity term known as *kinematic viscosity* (υ) expressed as

$$\upsilon = \frac{\mu}{\rho}$$

The unit of υ in SI units is m^2s^{-1} but commonly, the cgs unit of stokes (S or St = 10^4 m^2s^{-1}) or centistokes (cS or cSt) is used.

5.3 LAWS OF FLUID FLOW

Hydraulics concerns pressure, velocity and flow and discharge rates conforming to the laws of conservation of mass, momentum and energy. Important laws of hydraulics that are of use to fire professionals are discussed briefly in this section.

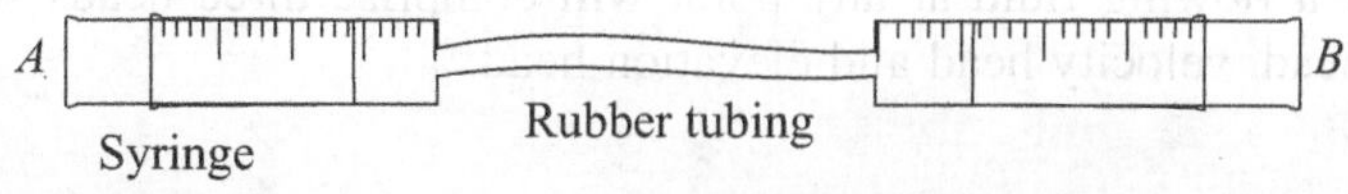

Figure 5.1 Demonstrating a principle of hydraulics.

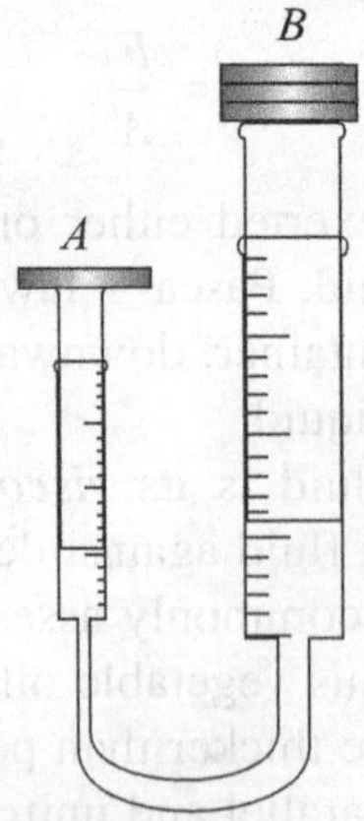

Figure 5.2 Uniform pressure distribution in fluids.

Some basic principles of hydraulics may be demonstrated by joining two syringes through a rubber tubing to make a watertight seal (Figure 5.1). The syringes can be filled with any fluid such as water, oil or just air. Pushing one of the plungers causes the other plunger to move out. This is observed even if the connecting tube is bent or the syringes are on different planes.

With syringes of different sizes, it is possible to demonstrate how to multiply force (Figure 5.2). A small mass on a small syringe can lift a much larger mass if it is resting on a bigger syringe (a principle applied in hydraulic jacks or car lifts). This is because the pressure in the two syringes is equal, but the force on each piston is different as their surface areas are different (Force = pressure × area).

5.3.1 Pascal's Law

It states that a static fluid exerts equal pressure (*isotropic*) on all its bounding surfaces in a direction perpendicular to the surface. Thus, in the case of water held in a pipeline closed at both ends to resemble a cylinder, equal pressure is exerted both on the walls and on the end faces. This pressure is called *static pressure* or *static head.*

If both the pipe ends are opened with one end connected to a water storage reservoir and the other end left open, water contained inside the pipe starts to flow and a new pressure is introduced. This pressure due to velocity of flow acts in a direction parallel to the pipe wall and is termed *velocity pressure* or *velocity head*. If the pipe wall is perforated at a point, the static head on the wall at that point converts to velocity head of the new branch stream. This change of energy from one form to another is in line with the law of conservation of energy.

When a source of energy is applied at an upstream point, flow is initiated. This source of energy may generally be either due to elevation or from a pump or a combination of the two and is called *elevation head.*

The energy of a flowing fluid at any point will comprise three heads we have discussed above viz., static head, velocity head and elevation head.

5.3.2 Newton's Law of Conservation of Energy

In Newton's law, the basic principle applicable in hydraulics is the ***law of conservation of mass***, also known as ***equation of continuity*** expressed as

$$\rho_1 A_1 v_1 = \rho_2 A_2 v_2 = \text{constant} \tag{5.1}$$

where, ρ is the density, A the cross-sectional area and v_1 and v_2 the velocity of the fluid at two points in the flow path. Physically it means that what flows in also flows out and, if the cross-sectional area of flow changes, the flow velocity will change to satisfy the equation of continuity. For incompressible fluids (e.g., water), $\rho_1 = \rho_2$ and the equation can be simplified to

$$A_1 v_1 = A_2 v_2 = \text{constant} = Q \tag{5.2}$$

5.3.3 Law of Conservation of Energy

The law of conservation of energy applied to incompressible fluids (like water) is expressed as the ***Bernoulli equation***. From the foregoing discussion, the equation can be written as

Pressure head + Velocity head + Elevation head = Constant

or
$$p + \frac{1}{2}\rho v^2 + \rho g h = \text{Constant} \tag{5.3}$$

where, p is the pressure head and gives the static pressure at a point, $1/2\,\rho v^2$ is the velocity head due to flow, and ρgh is the elevation head due to the level of the liquid. and,

ρ = density of the liquid
v = velocity of the flowing liquid
g = acceleration due to gravity, and
h = elevation above a datum level

The pressure and elevation heads comprise potential (stored) energy and the velocity head is the kinetic energy of a fluid system.

For a large overground reservoir open to atmosphere with an outlet near its bottom as shown in Figure 5.3, Bernoulli equation can be expressed as

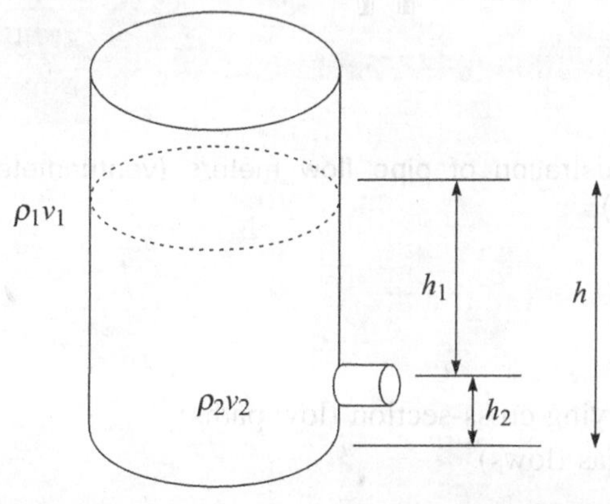

Figure 5.3 Overground reservoir open to atmosphere with outlet near its bottom.

$$p_1 + \frac{1}{2}\rho_1 v_1^2 + \rho_1 g h_1 = p_2 + \frac{1}{2}\rho_2 v_2^2 + \rho_2 g h_2 \quad (5.4)$$

In this case, $\rho_1 = \rho_2$, $p_1 = 0$ and $v_1 = 0$. Also as the orifice opens to atmosphere, $p_2 = 0$, equation (5.5) can be written as

$$g(h_1 - h_2) = \frac{1}{2} v_2^2 \quad (5.5)$$

If $h_2 = 0$ (outflow at datum level as in a hydrant), equation (5.5) becomes

$$v_2 = \sqrt{(2gh)} \quad (5.6)$$

And, volumetric flow rate Q is the product of cross-sectional area (A) and flow velocity. Thus,

$$Q = A\sqrt{(2gh)} \quad (5.7)$$

5.4 FLOW MEASUREMENT IN CLOSED SYSTEMS

To facilitate estimation of flow rates, different flow meters have been developed based on the principles and characteristics of fluid flow. Some of the common types of differential pressure flow meters widely used for closed pipe flow are (Figure 5.4):

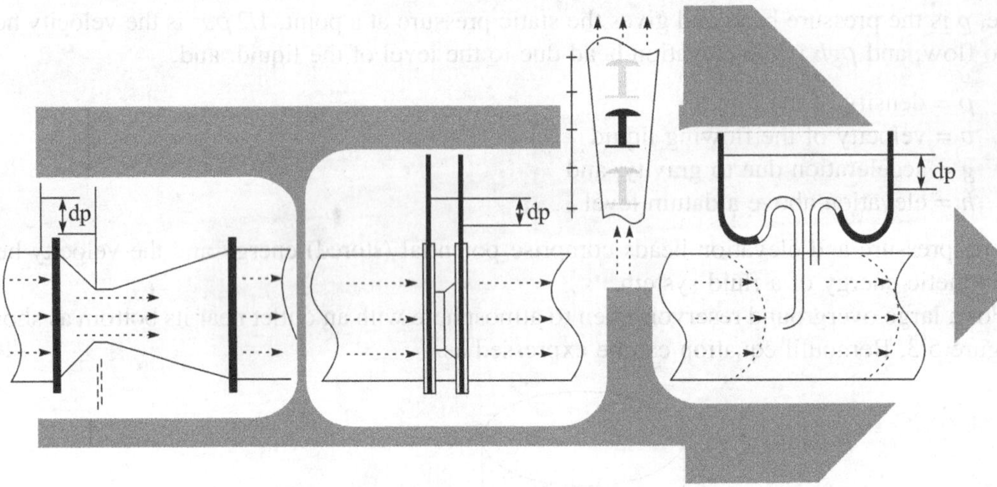

Figure 5.4 Composite illustration of pipe flow meters (venturimeter, orifice plate, rotameter and pitot tube).

- Venturi meters
- Orifice meters
- Pitot tubes
- Rotameters (for varying cross-section flow paths)
- Flow nozzles (for gas flows)

In differential pressure flow meters, a constriction is placed in the flow path to cause a pressure drop and this pressure differential is correlated to the flow rate using Bernoulli's theorem.

Some other types of flow meters used to measure flow rate in pipes include turbine meters, velocity meters and ultrasonic Doppler flow meters. Interested readers may refer to detailed descriptions in available literature.

5.4.1 Venturi Meter

It has a short reducing and then a long expanding pipe section. The reduction in cross section (constriction) causes a pressure drop and this is measured using a differential pressure measuring device like a manometer.

For horizontal flow ($h_1 = h_2$) of a non-compressible fluid ($\rho_1 = \rho_2$), from equations (5.4) and (5.5), we have

$$V_2 = \sqrt{\left[\frac{\{2A_1^2(p_2 - p_1)\}}{\{p(A_2^2 - A_1^2)\}}\right]} \tag{5.8}$$

and, volumetric flow rate ($Q = VA$) can be expressed as

$$Q = A_2\sqrt{\left[\frac{\{2A_1^2(p_2 - p_1)\}}{\{p(A_2^2 - A_1^2)\}}\right]} \tag{5.9}$$

5.4.2 Orifice Meter

It comprises a plate with a square-edged orifice placed across the flow path. The cross section of the flow stream as also the pressure are minimum at a point (called the *vena contracta*) downstream of the orifice plate. Accurate location of the vena contracta is vital for correct measurement of pressure drop between the orifice plate and the vena contracta.

Corrections have to be made in the equation for flow through an orifice to account for head loss and for the variation between the cross section of the orifice and actual cross section of the stream at the 'vena contracta'. Keeping details out, it can be said that the *discharge coefficient* C_d addresses the required correction and equation (5.6) is modified to

$$Q = A\,C_d\,\sqrt{(2gh)} \tag{5.10}$$

for flow through an orifice meter. The value of C_d depends on the orifice characteristics and flow pattern and vendors of orifice meters can furnish these to end users.

5.4.3 Rotameter

Rotameter is a flow measuring device installed in a vertical flow path of increasing cross section. It comprises a float which rises till the annulus between the tube and the float increases to maintain a state of dynamic equilibrium between the upward differential pressure and buoyancy against downward gravity. The position of the float indicates flow rate and the tube can be calibrated accordingly.

5.4.4 Pitot Tubes and Flow Nozzles

These are used more for pneumatic flows and their working principles are not discussed here.

5.5 FLOW MEASUREMENT IN OPEN CHANNELS

Rough estimates of flow rates in an open channel can be made by observing the velocity of a floating object on the surface of the channel and multiplying it with the cross-section area of the channel. Turbine flow meters can also be used to estimate flows in channels. A better method is to install overflow weirs, which are dam-like obstructions placed across the flow stream so that liquid flows over the top of the weir. Weirs notches can be rectangular or V-shaped (the latter, also known as triangular is preferred for low flow rates), covering the full width of the channel or a part of it and, provide either a sharp drop or a gradual one [Figure 5.5(a) and (b)].

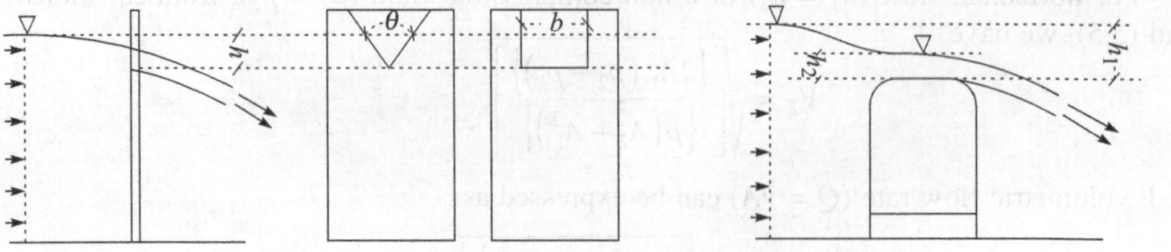

Figure 5.5 Types of weirs.

Flow rate over a rectangular weir is given by the equation:

$$q = \frac{2c_d b[2g]^{1/2} h^{3/2}}{3} \tag{5.11}$$

where,

q = flow rate
h = head on the weir
b = width of the weir
g = gravity
c_d = discharge constant for the weir (must be determined)

Manufacturers of standard weirs can provide c_d for their products as appropriate for measuring within specified head ranges.

For a *V*-notch weir the flow rate can be expressed as:

$$q = \frac{8c_d[2g]^{1/2}\tan(\theta/2)h^{5/2}}{15} \tag{5.12}$$

where, θ = *V*-notch angle

For broad-crested weirs, flow rate is expressed as

$$q = c_d h_2 b[2g(h_1 - h_2)]^{1/2} \tag{5.13}$$

5.6 FLOW AND FRICTION LOSS

As a basic hypothesis, fluid flow is assumed to be made up of thin layers (lamina) with a static layer adjacent to the wall (velocity = 0). Each inner layer moves relative to its adjacent outer layer and a velocity profile is set up through the mass of the fluid. In laminar flow, the velocity profile resembles a parabola with the least velocity at the wall surface and which steadily increases to maximum at the axis in case of a pipe. On the other hand, a flatter velocity profile represents turbulent flow [Figure 5.6(a) and (b)].

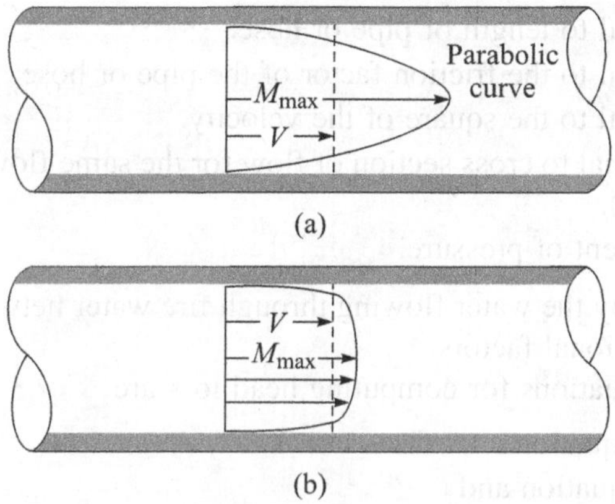

Figure 5.6 Velocity profiles of laminar and turbulent flows.

Without going into details, it may be said that an important dimensionless group called ***Reynolds number*** (Re) of a fluid flow determines whether the flow is laminar or turbulent.

$$\text{Re} = \frac{D_e V_\rho}{\mu} \tag{5.14}$$

where,

D_e = equivalent diameter = $4r_h$

r_h = hydraulic radius = $\dfrac{\text{cross sectional area of flow path}}{\text{perimeter of conduit}}$

V = flow velocity

ρ = specific gravity of the fluid

μ = the viscosity of the fluid

all in consistent units.

When Re ≤ 2100, flow is laminar and it is turbulent when Re ≥ 2300. For the intermediate range of Re, flow is a mix of laminar and turbulent. Frictional forces are more pronounced in turbulent flow than when flow is laminar. Water flows in fire protection systems will generally be in the turbulent domain.

5.6.1 Major Loss

Whenever a liquid flows — over a surface, in an open channel or through a pipe — there is head loss or pressure drop due to one or more of the following:

- Friction between the flowing liquid and containing surfaces.
- Friction among liquid particles including that due to turbulence.
- Change in flow direction.

- Abrupt decrease in cross-section of flow path.
- Abrupt change in flow velocity.

Energy lost due to friction in pipe flow is commonly referred as friction head loss or major loss. It may also be noted that frictional loss is:

- directly proportional to length of pipe or hose.
- directly proportional to the friction factor of the pipe or hose.
- directly proportional to the square of the velocity.
- inversely proportional to cross section of flow for the same flow velocity (one fifth of the flow path diameter).
- relatively independent of pressure.

Work has to be done by the water flowing through fire water networks or fire fighting hoses to overcome all these frictional factors.

Three widely used equations for computing head loss are

- Darcy–Weisbach equation
- Hazen–Williams equation and
- Manning's formula

Of these, the last viz., Manning's formula is applicable to flows in open channels and may not be very relevant for fire protection computations. Hazen–Williams equation is an empirical relation applicable only to water flows in the turbulent regime at temperatures up to approximately 25°C. A positive of the equation, however, is that the Reynolds number effects are included in it. The Darcy–Weisbach equation is applicable to different fluids but is dependent on Reynolds number. The Darcy–Weisbach friction factor for turbulent flows can be estimated using Moody's diagram available in standards texts.

In the Darcy–Weisbach equation, head loss during flow is given by

$$H_f = \frac{flV^2}{(2gd)} \tag{5.15}$$

where,

H_f = head loss due to friction (m)
f = Darcy–Weisbach friction factor
l = length of pipe (m)
V = velocity of flow (m/s)
d = diameter of pipe (m)
g = acceleration due to gravity (9.81 m/s^2)

Similarly, the Hazen–Williams equation expresses head loss in various forms such as shown in Equations (5.16) and (5.17)

$$h_L = \frac{V^{1.85}L}{(0.85C_H)^{1.85}R^{1.17}} \tag{5.16}$$

where,

h_L = frictional pressure drop, kPa/m
V = flow velocity, m/s
L = length of pipe or hose

R = hydraulic radius (Area/perimeter), mm
C_H = Hazen–Williams factor

$$\Delta P = 1.1101 \times 10^{10} \left(\frac{Q}{C}\right)^{1.85} \frac{1}{D^{4.87}} \tag{5.17}$$

where,

ΔP = frictional pressure drop, kPa/m
Q = flow rate, m^3/h
D = internal diameter of pipe, mm
C = Hazen–Williams C factor, (dimensionless)

The Hazen–Williams C factor depends on the type of the pipe material and the internal condition of the pipe. However, a value of $C = 100$ is used as a generally accepted average for many applications and, for pipes older than 10 years, $C = 75$ is considered to be more appropriate.

5.6.2 Minor Losses

Pipe fittings like valves, elbows and tees create additional losses over and above frictional losses and these are known as *minor losses,* although it can be significant and sometimes the term is a misnomer. For instance, in the case of short lengths of pipes with a large number of fittings, minor losses may exceed other losses.

In practice, *equivalent pipe lengths,* that is, the length of pipe that produces the same loss as the fittings (available from manufacturers and engineering data tables) are summed up for all fittings and added to the pipe length to arrive at a corrected length of pipe through which the water flows. This helps address the frictional losses due to pipe fittings installed on the system.

5.7 FLOW IN LOOPS AND GRIDS

When fluid flow is in loops or grids as is the case in hydrant, and other water-based fire protection systems, each node could be receiving inflow from different directions as also discharging in different directions. Loops and grids offer a big advantage in terms of pressure drops as well as alternate available routes of flow as compared to branched systems (This availability of alternate routes through grids is also a recommendation for access roads in the layout of a facility particularly to address emergency scenarios)

Where a loop involves a flow breaking into two branches and then recombining, flow in the branches will be equally distributed if the branches are symmetrical. Otherwise, the two branches are unequal with two unequal legs and the flows through the different legs will be different.

In practice, fire hydrant mains are generally laid in grids. On the other hand, where the end use involves sprinkler and water spray systems, risers etc., branched distribution pipes are in vogue.

5.8 HYDRAULICALLY REMOTEST POINT

Fire protection design standards routinely use the term ***hydraulically remotest*** (or ***most remote) point,*** which merits some discussion. A good explanation for the term is that it refers to that point in a closed flow system to reach in which the pump has to do maximum work. In a dead end horizontal pipeline, the end point (geographically most remote) is the hydraulically most remote point as, at that point, friction losses are the maximum. However, in case of a riser, an elevated point quite close to the pump geographically could be hydraulically remotest due to the vertical rise in flow path requiring the pump to do maximum work for water to be delivered at that point.

5.9 HEAD

With a specific duty in mind, users specify pumps on the basis of their discharge capacity and pressure. Designers, on the other hand, refer to discharge capacity and *head* — a term derived from Bernoulli theorem discussed earlier — which refers to the pressure exerted by a vertical column of a liquid due to its weight and is known as *static head.* For liquids with different densities, the same head would represent different pressures. The term is also used to express the kinetic energy developed by the pump. Thus, head is the height of a liquid column set up due to the kinetic energy imparted to the liquid by the pump.

5.10 WATER HAMMER

If high velocity of fluid flowing in a pipeline suddenly drops or if flow ceases completely, the kinetic energy of the fluid is passed on to the pipe walls or fittings. The effect of such pressure rise due to velocity changes in a water network or pipeline is known as *water hammer* and may crop up through sudden stoppage of a pump, closing of valves or even pipe rupture. If the pipe can absorb such shocks within its elastic deformation limits, there is no permanent damage but, if the effects go beyond elastic limits, pipe and fittings may break apart.

5.11 FRICTION LOSS IN FLOW THROUGH HOSES

Hoses are used in a range of fire protection applications. Non-collapsible suction hoses are used by fire appliances to draw water from sumps or ponds. Collapsible hoses may be used either to connect hydrant outlets to appliances or with different types of nozzles to apply water or foam on fires or for exposure protection. The aspect of friction loss is more relevant in case of connecting, attack or delivery hoses as they are used in one or more lengths of 15 m or 30 m each. Literature provides some data on friction loss per unit length of fire hose. However, friction loss coefficients used by fire fighters to estimate head loss in fire hoses were based on decades old hose types, while newer hose designs account for lower friction loss.

Based on the principles of friction loss outlined earlier in this chapter, a rule of thumb relation for friction loss for a hose is

$$P_f = \frac{9000 flQ^2}{d^5} \tag{5.18}$$

where,

P_f = pressure loss (bars)
f = friction factor
l = length of hose (m)
Q = flow rate (lpm)
d = hose diameter (mm)

Commonly, pipe or hose diameters are expressed in millimetres (mm). To convert head loss to pressure drop (bar), the following relation may be used:

$$P_f = H_f \tag{5.19}$$

5.12 PUMPS

A pump is a device that supplies energy to a liquid to increase its pressure and move it through a pipeline. Pumps have been classified in various ways such as in terms of their prime movers, their mode of operation, their design, etc. (Figure 5.7). Most firewater pumps in use today are of centrifugal type. However, positive displacement pumps are used as primers. Pumps may be considered the heart of a water-based fire protection system as they move water from a source to points of discharge or for application at desired rates and pressures (this is similar to the function of the heart as blood is delivered to all parts of the body). Gravity flow from elevated tanks alone will not be adequate to deliver the pressure required for most fire protection requirements and hence pumps are an essential component of many fire protection systems. Fire pumps may be fixed and installed in pump houses, they may be mounted on fire tenders and they may also be portable type that can be conveniently carried by two persons.

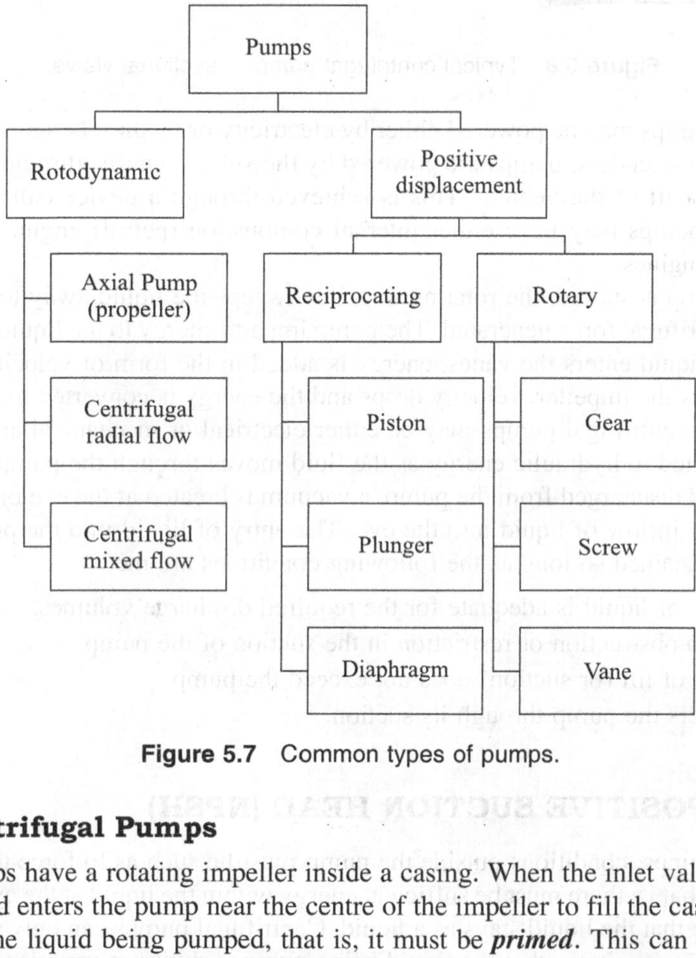

Figure 5.7 Common types of pumps.

5.12.1 Centrifugal Pumps

Centrifugal pumps have a rotating impeller inside a casing. When the inlet valve to the pump is opened, the liquid enters the pump near the centre of the impeller to fill the casing and immerse the impeller in the liquid being pumped, that is, it must be ***primed***. This can be ensured either by locating the pump at a lower elevation than the reservoir or by providing a check valve in the suction line to hold liquid in the pump, when the pump is idle. Centrifugal pumps normally have a valve on the discharge to control the flow and pressure (Figure 5.8).

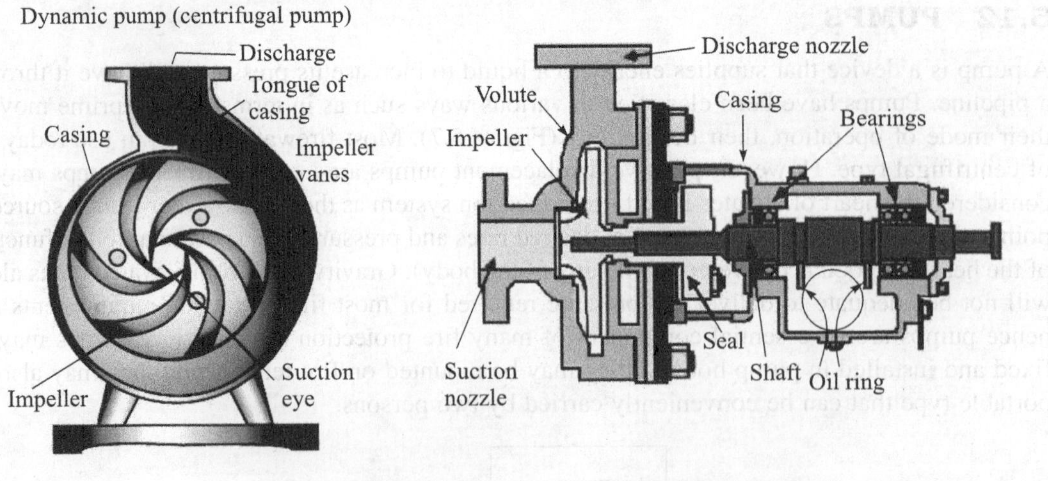

Figure 5.8 Typical centrifugal pump — sectional views.

Centrifugal pumps may be powered either by electricity or by diesel engines in case of fixed installations. On fire tenders, pumps are powered by the vehicle engine through a different shaft than the driving shaft of the vehicle. This is achieved through a device called *power take off* (PTO). Portable pumps may have either internal combustion (petrol) engines or compression ignition (diesel) engines.

When the pump is started, the rotating impeller sweeps the liquid away to the periphery of the casing by centrifugal force generated. The pump imparts energy to the liquid handled through the impeller. As liquid enters the vanes, energy is added in the form of velocity. Subsequently, as the liquid leaves the impeller, velocity drops and the energy is converted to pressure or head. Prime movers for centrifugal pumps may be either electrical or mechanical and these forms of energy are converted to hydraulic energy as the fluid moves through the pump.

As the fluid is discharged from the pump, a vacuum is created at the eye of the impeller and this causes further inflow of liquid into the eye. The entry of liquid into the pump followed by its discharge is sustained so long as the following conditions are met:

- The supply of liquid is adequate for the required discharge volumes.
- There is no obstruction or restriction in the suction of the pump.
- The height of lift (or suction) does not exceed the pump.
- No air enters the pump through its suction.

5.13 NET POSITIVE SUCTION HEAD (NPSH)

For centrifugal pumps, conditions outside the pump must be such as to force the liquid into the eye of the pump, that is, there must be sufficient energy within the liquid at the eye to perform this and also to ensure that the liquid stays as a liquid. Centrifugal pumps can only pump liquids and not vapours. This necessitates that the liquid being pumped does not vaporise at any stage — a situation that may arise with rise in temperature or drop in pressure. Vaporisation begins when the vapour pressure of the liquid equals atmospheric pressure. Thus the pump needs to have adequate suction head to prevent vaporisation at the lowest point in the pump. Net positive suction head (NPSH) is a measure to prevent liquid vaporisation or the energy in the liquid at the centre line of the pump to prevent vaporisation.

This energy is called *net positive suction head* (NPSH). It can also be described as the head due to which the liquid reaches the eye of the impeller passing through the suction pipe and fittings en route. If the liquid source is above the pump, the suction head is the sum of atmospheric pressure and static head. Thus,

NPSH = (atmospheric pressure) + (static head on suction)
– (intake line losses) – (vapour pressure of the liquid)

If, on the other hand, the pump is above the source, the suction head is the difference between atmospheric pressure and static 'lift'.

NPSH = (atmospheric pressure) – (static head on suction)
– (intake line losses) – (vapour pressure of the liquid)

NPSH is also defined as the difference between net suction head and vapour pressure at a rated flow or as the energy required to overcome intake head losses and fill the pump on the intake side.

NPSH are of two types: Required NPSH and Available NPSH,

Required NPSH is a function of pump design and is determined by the manufacturer through actual tests. As the liquid enters the eye of the impeller and moves on to the tip of the vanes, there are various pressure drops. Required NPSH is the head required such that these various pressure drops will not cause vaporisation and consequent vapour accumulation within the pump. The test results are provided by the manufacturer as the performance curve which correlates head with flow. Required NPSH is independent of the liquid density.

Available NPSH, on the other hand, is a function of the system in which the pump operates and is the excess pressure of the liquid in metres over its vapour pressure at the pump suction.

For a pump to perform its desired function, the available NPSH must be equal to or greater than the required NPSH. Based on the required NPSH given by the vendor for a pump, available NPSH can be calculated using Bernoulli equation at the pump centreline as

$$\text{NPSH}_{\text{av}} = \left(\frac{p_a}{\rho g}\right) + z - h_L - \left(\frac{p_v}{\rho g}\right) \qquad (5.20)$$

where,

p_a = absolute pressure
ρ = density of the liquid
g = acceleration due to gravity
z = elevation of pump to ensure $\text{NPSH}_{\text{av}} > \text{NPSH}_{\text{req}}$
h_L = friction head loss in intake line and
p_v = vapour pressure of the liquid.

The correlations among head, capacity, power and efficiency for any pump (known as ***characteristics*** of the pump) are expressed in the form of pump characteristic curves. These curves are used to select a pump and to determine its adequacy for specific duties. Pump manufacturers generally have standard sizes of pumps in which impellers of different sizes can be fitted to meet customer requirements.

5.14 CHARACTERISTICS CURVES

Positive displacement pumps deliver a definite volume of a fluid and in an ideal pump, the speed at which it operates is the only factor that effects flow rate. Figure 5.9 shows the characteristic curve and actual performance of a positive displacement pump.

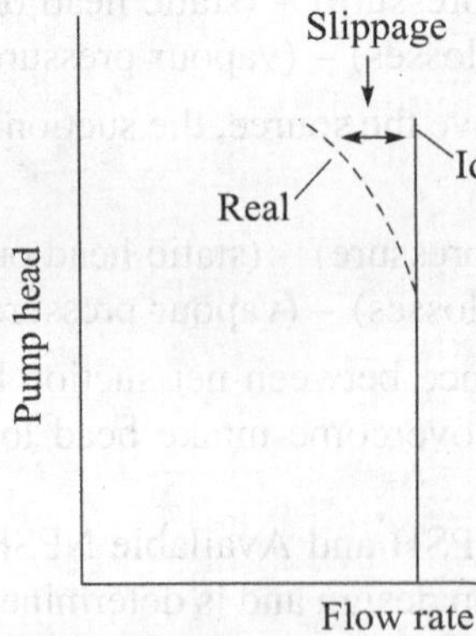

Figure 5.9 Characteristic curves for a positive displacement pump. "Standard graph".

Centrifugal pumps with a relatively flat characteristic curve are seen to be appropriate for use as firewater pumps as such a curve represents steady, non-pulsating flow of water at a uniform pressure over a wide range of flow (Figure 5.10). Such pumps can also run for some time with the discharge valve shut without damaging pump parts.

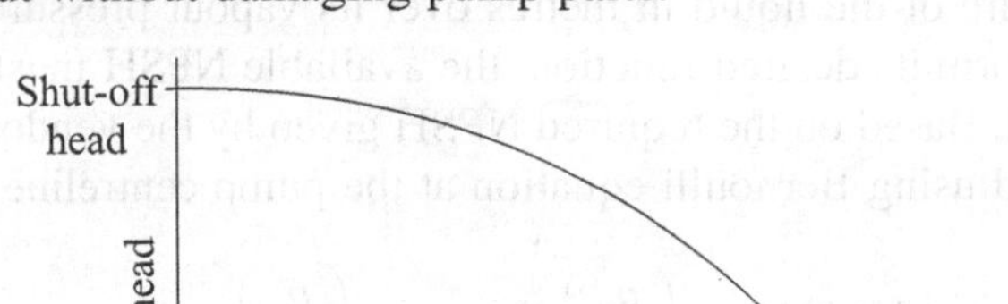

Figure 5.10 Characteristic curves for a centrifugal pump. "Standard graph".

5.15 PUMP COMBINATIONS

If increased flow rates are desired, multiple centrifugal pumps are connected in parallel; if on the other hand, higher discharge heads are desired, pumps may be connected in series, that is, the discharge from any one pump is fed to the suction of a subsequent pump. The latter may also be achieved by using multistage pumps where several impellers are mounted on the same shaft and where the discharge from each impeller enters the suction of the next impeller and each impeller provides an incremental rise in pressure. Firewater systems, generally, operate on relatively low pressures (around 1000 kPa) and single-stage pumps are adequate. Special cases like offshore services where high lift is required from the water source, multistage pumps may be required.

Centrifugal firewater pumps may be single or multistage with horizontal or vertical shaft with the orientation of the shaft describing the pump, e.g., a vertical turbine pump may have an electric motor positioned vertically on the pump shaft or be driven by a right angle gear drive to a horizontal diesel engine. Centrifugal pumps for firewater service are fed under positive pressure from aboveground supply sources or are submerged in the water supply such as in a well or sump.

5.16 CAVITATION

The phenomenon of ***cavitation*** occurs when the pressure of a liquid is reduced below its vapour pressure and then raised back above its vapour pressure. Bubbles of vapour form at first and these bubbles collapse when higher pressure condition is reached. The collapse occurs at sonic speed ejecting minute jets at very high velocity. These jets can cause severe erosion when they impinge on even the hardest materials over a long time.

Cavitation is likely if the net suction head is less than the vapour pressure of the liquid being handled and the liquid can vaporise inside the pump. As the bubbles formed move from the central low-pressure region of the pump to high-pressure areas near the tip of the impeller vanes, they collapse violently. This leads to localised stresses on the impeller or the pump casing that may cause vibrations and lead to their mechanical damage. Some of the causes that may lead to cavitation in fire water pumps are

- Restriction at inlet
- Collapsed suction hose
- High water temperature
- Suction hose small for desired volume discharge

Cavitation can be indicated by rattling noises in the pump and increased vibration. In vehicle-mounted pumps, increased engine speed without corresponding rise in discharge pressure may also be observed. In such case, the pump throttle should be reduced and other corrective actions initiated.

5.17 PUMP SELECTION

Fire protection by implication concerns storage, transport and delivery of vast quantities of water in various ways to desired locations. Selection of the appropriate pumps is, therefore, of primary importance. In India, comprehensive guidelines for selecting pumps and determining adequacy are stipulated by the Tariff Advisory Committee (TAC). Some of the major points made by TAC are:

- Pumping capacity based on requirement of various protection systems like hydrants, foam system, sprinkler and spray systems.

- Duel power source (electric and diesel).
- At least two pumps of equal capacity on different power.
- 150% rated capacity at 65% of rated head.
- Shut-off pressure.

The following factors are known to aid economical selection of pumps.

- Maximum discharge under the most demanding design conditions.
- Total head-capacity relation (characteristic curve).
- Available NPSH.
- Pump speed and power source requirements.
- Space and environmental requirements.
- Maximum allowable system head downstream of pump discharge.
- Independent power source or diesel driven. Immediate start and discharge.

5.18 PRIMERS

Centrifugal pumps cannot pump air — they need to be filled with the liquid it is meant to handle. ***Priming*** is the process of removing the air present inside a pump casing and suction hose such that atmospheric pressure can force the liquid up to the pump casing. Given below are different methods of priming:

- If the water supply is from a higher elevation than the pump, then water flows into the pump by gravity pushing the air out.
- When the water is drawn from a hydrant, mains pressure is sufficient to force water into the pump pushing the air out. (*Never use a primer in this case*).
- If the water supply is below the pump, an extra piece of equipment called a ***primer*** is required. Its job is to remove the air from inside the pump and lowering the pressure so that atmospheric pressure can force water into the pump (Remember, fluids will flow from high pressure to low pressure).

When air from inside the suction hose is removed by the primer, atmospheric pressure pushes water up the hose and into the eye of the pump. The completion of the discharge of air is indicated when water is released through a relief point. The distance or height that the water can be moved by atmospheric pressure is theoretically 10 m. In practice, however, it is closer to 8 m. Primers are very useful on appliance-mounted pumps unless the pumps are 'self-priming' type.

5.18.1 Types of Primers

Primers themselves are types of pumps and three common types are:

- Reciprocating
- Exhaust ejector
- Water ring

A ***reciprocating primer*** consists of a small piston pump connected to the suction side of the main pump through a pipe with a valve. This pump is driven by the main pump shaft. Before starting this pump, the relief air valve is opened and the pump drive engaged. The downward stroke of the pump reduces pressure inside he cylinder and its upward stroke forces the air out. After several strokes, the air inside is completely removed and water from the open source enters the pump casing. The completion of priming is indicated by water flowing out of the waste pipe.

In an ***exhaust ejector primer***, the vehicle engine exhaust is diverted from its normal route to a nozzle, and discharged at a high velocity through a throat pipe. This in turn draws in air from inside the system by venturi action and discharges through a relief. When priming is achieved water flows out. The efficiency of the exhaust gas primer depends on the velocity of the gases leaving the discharge nozzle and hence priming is done at high speed.

A ***water ring primer*** has an impeller with a hollow centre in an elliptical housing. A stationery boss inside the hollow centre projects out of the housing cover and has two ports in the periphery for suction end. The housing is filled with water before priming. Centrifugal force due to a spinning impeller forces the water in the housing to move outwards and form a ring of water at the periphery and create a hollow elliptical vortex. The diameter of the water ring changes, increases and decreases with each revolution. This creates a pumping action and air from the casing is pushed out as water enters the suction. This primer may be automatic or manually operated. Power for the primer is taken either from the main shaft or from a friction wheel.

5.19 EDUCTORS

Besides pumps, some other means of transferring liquids are also used in fire protection systems. For instance, water streams are taken through venturi-type devices where the drop in pressure at the constriction is utilised to draw in foam compound and make foam solution for subsequent application as expanded foam. These devices are called *eductors*.

5.20 CONCLUSION

Water continues to be a major element of fire protection systems and perhaps will continue to be so in the foreseeable future. Hence, it is vital for fire protection professionals to have a good understanding of hydraulics and equipment and appliances associated with fire protection. Some applications of water in fire protection systems are described in Chapter 9.

QUESTIONS FOR REVISION

1. Explain the significance of the terms *head* and *discharge pressure* delivered by a centrifugal pump as also of their units of measurement.
2. With reference to Bernoulli equation, explain the outcome of perforating the mains of a pressurised fire hydrant network at a point.
3. What are the conditions of flow under which different head loss equations are appropriate in closed or open flow systems?

4. Describe the parts of a centrifugal pump and explain its working principles. What are the different sources of energy that can be used to operate a centrifugal pump?
5. What is the purpose of a primer? Describe the working principles of different types of primers. At what locations are primers not required?
6. Water is known to be flowing at a velocity of 2 m/s through a 60 mm diameter canvas hose connected to a hydrant. Using a friction factor of 0.007, find the pressure lost due to friction in 100 m of the hose.

6
Fire Science for the Built Environment

Doctors bury their mistakes, and lawyers get to lock theirs away;
but engineers build their mistakes out in the open for everyone to come and see.

—An Engineering Professor

6.1 INTRODUCTION

Some analysts believe that the collapse of the New York World Trade Centre towers on September 11, 2001 was not only due to the impact of the striking aircraft but also the effects of the ensuing fire that led to collapse of the landmark buildings. These buildings were not designed for the type of fire loads generated within the buildings that day. Not a surprise though because fire safety provisions in building codes are generally not intended to cater for such eventualities.

Mangled steel, peeled cement plaster, crumbled brick walls, shattered glass and charred wood are sights of post-fire scenarios that many of us have seen or known. While many materials used in the construction of buildings and industrial structures may not be fuel for a fire, all materials succumb to the effects of sustained fire conditions exceptions being refractory like materials that are designed for use in high temperature conditions.

6.2 BUILDING LOADS

Buildings are subject to various loads which may be broadly classified as:

- ***Dead load*** which is a built-in, permanent feature comprising the structural and non-structural components of the building.
- ***Live load*** comprises loads that are not attached or permanent and may be considered as load due the contents and occupants.
- ***Other loads*** may be made up of impact load due to activities within the building, wind loads, loads due to earthquake, etc.

The above loads are all forces and are expressed in units of force.

Fire Load, on the other hand, is the potential heat release rate (energy per unit area) that will be released if all combustible building components and contents are completely burnt. Thus the unit of fire load is J/m^2. Sometimes, however, a hypothetical fuel of known heat of combustion is considered such that fire load may be expressed in terms of load of fuel per unit area.

Fire load, on the other hand, is described as the heat of combustion that may be released if all combustible contents in a compartment are completely combusted divided by the floor space area of the compartment.

A significant and interesting learning from the above is that a warehouse storing tissue paper may have a low structural load but high fire load while, if the same warehouse is used to store steel, the structural load will be high and the fire load will be very low.

6.3 LIFE SAFETY AND PROPERTY PROTECTION

The purpose of a structure is to permit safe occupancy for people and the designed activities. This same purpose is enhanced through any fire protection measure with overriding priority for life safety. Hence, in most countries, statutory requirements for fire protection are based on life safety concerns and property protection features are intended to limit physical damage. Generally, these objectives are complementary where the primary concern is with the conservation of property, its function or contents, the management of the asset can implement measures beyond those limited to address life-safety concerns.

6.4 MATERIALS OF CONSTRUCTION

Most materials of construction such as bricks, steel, cement, concrete or timber are composites made up of different materials that either occur naturally or are brought together through different processes. Their behaviour in fire conditions vary depending on many factors like their constituents, their material structures, the manner in which they are used and, in the case of man-made materials, the process they have undergone to make them fit for construction.

Building materials have been classified based on their performance under fire conditions as shown below:

1. Those that are used for their stress (usually tensile) carrying properties, e.g., steel in different forms.
2. Those that can carry moderate stress (mainly compressive and thermal) and provide insulation to the above, e.g., brick and concrete.
3. Insulators that cannot carry loads, e.g., gypsum board.
4. Insulators that can carry loads, but are also combustible, e.g., timber.
5. Insulators that cannot carry loads and are combustible, e.g., wood laminates.

6.5 FIRE PROPERTIES OF MATERIALS

When involved in a fire, different materials of construction and furnishing do not only succumb to the prevailing conditions of high temperatures, some also release harmful products of combustion.

Let us take a look at the fire behaviour of some of the common materials used in residential, commercial or other buildings.

6.5.1 Timber/Wood

Till the advent of the steel and concrete, timber was used for most load bearing structures in buildings. In many parts of the world, wood continues to be an important material used in building frameworks. With the reduction in forest cover and the resultant rise in the cost of timber and wood-based products as also due to the shift towards high-rise structures, timber has largely been replaced by other structural materials today. However, wood and wood-based products like plywood and particle boards are used in doors, wall and ceiling panels and furniture primarily for aesthetics and its adaptability.

Wood is a matrix of complex cellulose with moisture and air trapped within. Burning properties of wood depend on species, its shape, manner of use and any type of treatment applied. When exposed to sustained high temperatures, thick wood will char, which acts as a shield for the inner material (some other solids like PVC also display char formation) but processed materials like plywood disintegrate at high temperature as adhesives used lose their binding property.

For flaming fire to occur, wood must be heated till it pyrolyses and releases vapours that mix with air to form combustible vapour–air mixtures. Heat energy required for ignition by pyrolysis depends on factors like species, size, orientation, texture, shape, moisture content, etc.

Wood is a bad conductor of heat and this property has traditionally been used in making 'fire doors' by placing wood between sheets of mild steel. Among methods developed to protect wooden members from the effects of fire are immersing sections in fire resistant solutions of metal salts and fire retardant surface coatings. The burning of wood releases hydrocarbons and vapours like toxic aldehydes, besides common products of combustion.

6.5.2 Steel

Steel is possibly the most used structural material today either by itself or in combination with other materials like concrete. Exposure to high temperatures as in fire conditions can bring about significant effects like expansion, change in microstructure and lowered elastic and yield strengths. Small cross-section steel items such as wires, strips or thin rods exposed to high temperatures may exhibit similar change in properties across the cross-section but in larger structural steel members, temperature profiles may be set up across the length and breadth of the member till sustained exposure induces some equalisation. It is important to recognise and remember that high skin temperature alone of a structural steel member will not determine the likelihood of failure as a range of factors such as dimensions, duties and duration of exposure will play a role.

Thus, under fire conditions, expansion on the periphery of an exposed steel member will be greater than in inner sections. In some cases, a steel member may be heated on one face while the opposite face may be relatively cold causing uneven expansion and bending. Linear expansion of a beam can impart stresses on its supporting columns while bending can affect the stability of the floor or roof it supports. Major expansion or bending can impart excessive stresses on the entire structure and cause it to fail. This is seen to happen in major fires in buildings or in open industrial plants where exposed steel structures were used.

In industry, a variety of steels (mild steel, high carbon steel and stainless steels) are used for structures, process equipment and vessels, storage vessels, machinery and other purposes. Structures may generally have limited built-in fire protection while in process vessels used in chemical process industries, external insulation may offer some degree of fire resistance. Learning from loss experience, however, design codes recommend load bearing steel structures to be made fire resistant at least at lower levels by encasing in concrete or other forms of thermal resistance. Active fire protection measures like water spray systems are also widely used to protect steel against the thermal effects of fire.

6.5.3 Cement

The process of manufacturing cement from raw materials such as limestone and steel plant slag involves high temperatures and, hence by itself, may not be susceptible to fire. However, cement is commonly used in combination with sand and water as binding agent for bricks or for surface finishing. Cement plaster made of sand, cement and water aggregate tends to disintegrate when exposed to fire conditions cause loss of moisture. Vermiculite (mica based) cement plaster has been used as an effective fire shield for steel. The other and very important use is in the form of concrete that may contain cement, sand and stones giving a porous matrix with entrapped air and moisture.

6.5.4 Concrete

Concrete in its various forms is a most preferred construction material. Some of the water used in making concrete aggregate is retained within the set matrix, the loss of which in fire conditions causes 'spalling' whereby the concrete develops cracks. Spalling may expose the steel inside to overstressed conditions and lower its load bearing capacity. Under fire conditions, the exposed and unexposed sides of a concrete member may be at very different temperatures leading to uneven expansion and unequal dehydration at two opposite surfaces. High temperatures within the matrix may also cause expansion of the reinforcing steel in prestressed concrete.

Concrete is a good thermal insulator and hence is used to protect steel structures. In fact, design standards recommend that supports for storage and process vessels be encased in concrete to provide a degree of fire protection to the supports.

6.5.5 Brick

Bricks can be of a wide variety ranging from different types of bricks used in building construction to refractory bricks meant for use in very high temperature environments in industry. Traditional bricks used in building construction are made of burnt clay. Having already gone through fire, bricks can withstand fire conditions without damage but the cement plaster joining the bricks can give way causing brick structures to collapse. Being good insulators, bricks are used to erect fire resistant walls or partitions.

Hollow cement bricks provide a degree of insulation due to the air space inside and may resist heat transfer from one side to the other but can spall and collapse.

6.5.6 Glass

Glass is used mainly to admit light and permit vision, or as a decorative and finishing material. Some modern commercial buildings have entire external walls made of glass. Different types of glass have been developed by altering either the materials or the process of manufacture. Irrespective of the type, glass is a bad conductor of heat although the degree of thermal conductivity may be different. This property of glass is well-known as seen from cracking of tumblers of thick glass when hot water is poured into them. In building fires, plain glass window panes are seen to crack due to the same property. Some varieties of glass can also soften at relatively low temperatures and then melt. To provide some fire resistance to glass, embedding wire mesh helps.

When glass is used on walls and floors are conventional cast concrete, gaps are left between the floor and the wall. This negates any intended compartmentation and enhances the potential of fire spread through these gaps.

6.5.7 Some other Materials

Aluminium is finding increasing use in buildings today. The metal softens at relatively low temperature and easily succumbs to fire conditions.

Plastics have entered every field of life. In buildings, polyvinyl chloride (PVC) may be used for doors, partitions or plumbing and polystyrene for thermal insulation. In addition, many different plastics may be present in a building as furniture, cushions, household appliances or electronic goods. All plastics will succumb to fire conditions, although the route to their destruction may be different. A very dangerous aspect of plastics is the toxic character of many of the products of combustion.

When heated, PVC decomposes to release chlorine atoms, which can combine with hydrogen to give toxic and corrosive hydrogen chloride (HCl). Highly toxic phosgene ($COCl_2$) is produced in negligible amounts from the combustion of PVC. Carbon monoxide (CO) may also be produced when PVC is involved in a fire. PVC is not readily flammable but will burn freely in temperature conditions high enough to bring the plasticisers into a volatile state. Burning PVC produces copious quantities of corrosive and toxic heavier-than-air gases that tend to settle at lower levels. Heavy black smoke, mostly consisting of carbon particles, is also given off which can affect electronic and electrical equipment in the area. There is evidence that PVC smoke can contaminate and de-sensitise ionisation type smoke detectors.

Polystyrene is used extensively in packaging and insulation. When heated, the polymer is decomposed and the major product is styrene monomer (toxic), which is further broken down to smaller molecules that react with oxygen to form the usual combustion products and CO.

Gypsum board is used extensively as wall and ceiling panels. While it may not itself be combustible, gypsum boards can disintegrate when exposed to high temperatures.

6.5.8 Building Contents

While materials of constructions may not be many, the contents of built facilities are of a very wide variety and most can contribute fuel to a fire. Depending on the type of occupancy, contents could range from those commonly present in living premises to all types of hazardous materials that are stored or processed. Thus, besides contributing to fire load under fire conditions, contents of built premises can pose health hazards too.

The time–temperature curve for a location depends primarily on the contents and the initial temperature rise after ignition will expectedly be very rapid for hydrocarbons as compared to a normal residential or office compartment. The demands on the structural and non-structural members of the building will be very different in this case.

6.6 FIRE TESTING

To design and construct buildings that are safe for people to be in and also to exit in a safe way in case of an emergency, materials and building components need to be tested and proven for their fire behaviour. Fire testing can be divided into two distinctive areas: ***resistance-to-fire*** and ***reaction-to-fire***. Resistance-to-fire testing can be used to qualify full size construction elements/ assemblies and service installations for their ability to contain a fire or protect an area from fire such as fire-rated doors, walls, partitions, dampers, etc., for a period of time. On the other hand, reaction-to-fire testing can be used for any single surface materials and coatings to determine combustibility, contribution to fire, smoke generation and rate at which flames spread.

6.7 TESTING FOR RESISTANCE-TO-FIRE

The purpose behind resistance-to-fire testing is to determine for how long the assembly will hold up in a fire scenario in terms of the following three properties:

- Integrity (does not break up) and lead to flame penetration.
- Insulation (does not allow high temperature rise on the unexposed face).
- Stability (continues to perform its design intent).

Based on these principles, the European standard EN 13501–2 specifies the criteria to characterise the fire resistance of a building system:

R Load-bearing capacity (the ability of the building element to maintain strength and stability of the building system over a period of time).

E Integrity (the ability of the building element to remain intact, even if the surface temperature is at a critical level).

I Insulation (the ability to keep the temperature lower than 140°C on the unexposed side), expressed in minutes.

A product fulfilling all these basic criteria for 90 minutes will be classified REI 90.

It should however be recognised that the 'standard fire' does not represent heat transfer to a structure and that it is only a standardised screening method.

6.8 STANDARD TIME–TEMPERATURE CURVES

Resistance-to-fire or simply fire resistance tests are conducted as stipulated in various standards. Fires may generally be considered in two categories viz., ***cellulosic*** and ***hydrocarbon***. Cellulosic fires are those that involve cellulosic materials (timber, fabrics, paper, etc.) as may occur in buildings like homes, offices, hospitals, shopping malls, schools, etc., and hydrocarbon fires are those where the fuel may be gas, petroleum, petrochemicals, etc. The difference between the two types is not the temperature reached but the time taken by the fire to reach its maximum temperature range. Under test conditions, a hydrocarbon fire will reach a temperature of 900ºC in 8 minutes, while a cellulosic fire will take 60 minutes to reach the same level (Figure 6.1). However, fire tests developed are not limited only to these two types and these as well as some others are outlined below:

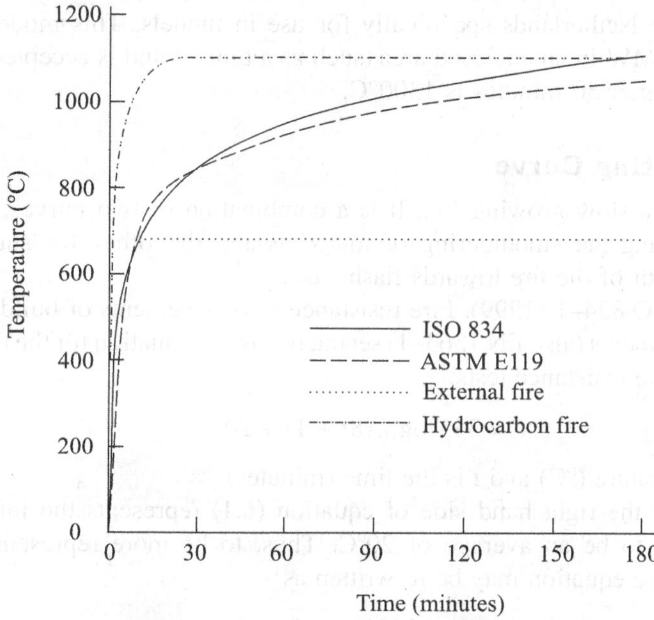

Figure 6.1 Standard time–temperature curves.

6.8.1 Standard Cellulosic Time–Temperature Curve

This is also known as the 'A' curve. This curve is used in standards throughout the world, including ISO 834, ASTM E-119 (USA), BS 476-Part 20 (UK) and EN 1363-1 (European Norm). It is a model of a ventilation-controlled natural fire, i.e., fires in a normal building. The temperature rise after 30 minutes is 842°C.

6.8.2 Hydrocarbon Curve

This curve (also known as the 'H' curve) simulates fire in a hydrocarbon pool in the open and is used on load-bearing steel structures such as legs supporting vessels or columns or structures supporting pipe racks. The curve is not meant for use on items like process vessels per se. The temperature rise after 30 minutes is 1110°C.

Although the 'H' curve is based on a standardised type fire, in the real world there are numerous potential types of fire in the hydrocarbon sector with a wide range of fuels and durations ranging from very brief to days.

6.8.3 External Fire Exposure Curve

This time–temperature profile is meant to represent exposure of a building to fire in an adjacent building or to an open fire outside. In such a case, there are different ways by which heat from the fire may be dissipated and hence, a reduced level of heat exposure. The temperature rises approximately 680°C after 20 minutes and then remains constant.

6.8.4 Rijkswaterstaat (RWS) Curve

It is developed in the Netherlands specifically for use in tunnels. This model is based on a petroleum fire of 300 MW in an enclosed area (such as a tunnel) and is accepted internationally. The temperature rise after 30 minutes is 1300°C.

6.8.5 Slow Heating Curve

This profile simulates a slow growing fire. It is a combination of two curves, one for the first 21 minutes representing the smouldering of materials and the other for subsequent periods representing the growth of the fire towards flashover.

IS 3809–1979, ISO 834–1 (1999), Fire resistance tests – elements of building construction, Part 1: general requirements (also EN 1363–1) set the following equation for the time–temperature profile in cellulosic fire resistance tests:

$$T = 345\log_{10}(8t + 1) + 20 \tag{6.1}$$

where T is the temperature (°C) and t is the time (minutes).

The figure 20 on the right hand side of equation (6.1) represents the initial and ambient temperature assumed to be an average of 20°C. Thus, to be more representative of various ambient conditions, the equation may be re-written as

$$T = T_o + 345\log_{10}(8t + 1) \tag{6.2}$$

where T_o is the initial ambient temperature at the test site.

The hydrocarbon curve as described in EN 1363–2 Fire resistance tests – Part 2: Alternative and additional procedures conforms to the following profile:

$$T = 1080(1 - 0.325e^{-1.067t} - 0.675e^{-2.5t}) + 20 \tag{6.3}$$

As in the case of the cellulosic equation, equation (6.3) may be rewritten as

$$T = T_o + 1080(1 - 0.325e^{-1.067t} - 0.675e^{-2.5t}) \tag{6.4}$$

In different test laboratories of the world including the one at the Fire Research Centre of the Central Building Research Institute (CBRI) at Roorkee, gas fired furnaces are programmed to replicate the standard time–temperature curves and structural components loaded to their intended design stresses are exposed to these conditions inside the furnace and the time to failure observed which is then declared as its fire resistance. Resistance-to-fire tests on beams and columns are conducted in the form and loaded the actual intended way, wall sections and fire doors are tested as vertical sections, while roof/ceiling or floor/ceiling sections are tested as horizontal sections.

As the tests take much time and effort to be conducted besides being expensive, analytical techniques have been developed to replicate test conditions. These techniques include radiant heat transfer for exposure, conduction for internal heat transfer and convection to address the existence of voids within the structure of the member.

The shape and dimensions of structural members are important in fire resistance as the heat has to be transmitted inwards till damaging temperatures reach a critical distance into the material matrix. For this to occur, fire conditions must persist, as momentary exposure to even very high temperatures may not lead to failure of structural members. It is for this reason that the term *fire resistant* rather than *fireproof* is preferred to describe the ability of a material to withstand fire conditions.

6.9 OTHER FIRE TESTS

The resistance-to-fire tests are meant to assess the capability of structural components to continue to perform their designed duties even in other fire conditions. There are other aspects of building materials and the contents in the building which can aid the growth and spread of fire. These include properties like ignitability, flame spread and smoke generation, which may, to varying extents, be significant for fire spread as well as life safety. Thus the fire properties of wall and ceiling panels, flooring and floor coverings, cables, drapes, furniture, upholstery and other contents gain significance.

Further, fire properties of a material are also affected by its orientation. Recall that in a campfire, the inclined orientation of pieces of firewood makes it easier to light them and they burn better too. In a wood stove, the arrangement of the firewood is important. Similarly, a plywood panel will behave differently in fire conditions depending on whether it is mounted on the wall or on the ceiling.

An important reaction-to-fire test is the (Steiner) Tunnel test named after the designer of the test at Underwriters' Laboratories Inc. In this test, known as the Method of Test of Surface Burning Characteristics of Building Materials (ASTM Standard E-84), a sample of material 7500 mm × 450 mm is installed on the underside of the removable top panel of the test chamber (tunnel). A constant, regulated draft is maintained in the tunnel and a gas flame is applied at the upwind end. The advancement of the flame front along the length of the sample is observed through side windows and timed. It is then compared against the two standards viz., asbestos-cement board (rated zero) and red oak (rated 100) and a number given which provides a relative rate of flame spread over the material surface. It may be noted that the test result is not indicative of fire resistance of the material. NFPA 101 – The Life Safety Code (of the National Fire Protection Association, USA) groups flame spread ratings into three classes: A (lowest) to C (highest).

6.10 PASSIVE FIRE PROTECTION

Built facilities may be provided fire protection either by building in fire safety measures or by means of early control through detection and extinction methods. Built-in measures introduced through resistance-to-fire and reaction-to-fire improvement provisions may be classified as

passive fire protection, while the latter may be termed ***active fire protection***. Active fire protection systems function by reacting to conditions caused by a fire such as heat, smoke or light to impede the fire process. Active and passive fire safety provisions are complementary, not competitive.

The concept of passive fire protection is based on the principles of:

- Controlling likelihood of ignition.
- Containment of fire and its effects within an area — compartmentation, opening protection like fire doors and wired glass, fire stops, dampers, etc.
- Conferring fire resistance to the structure and building elements for a specified time period — structural fire protection achieved with fireproofing materials such as sprayed film intumescents, gypsum-based plasters and cementitious products, mineral wool wraps and insulation, cladding or encasing in concrete.
- Enhancing reaction-to-fire behaviour of exposed surfaces such as walls, ceiling, flooring - coating, treatment, cladding, etc.

With passive fire safety in mind, in the case of a building compartment, it is so constructed that, should a fire occur in the compartment, it will be restricted to that compartment for a period of time. Doors and other similar opening covers in compartment walls should be able to withstand the effects of fire for the same period of time as the walls themselves. Fire dampers should be installed in ducts where they pass through compartment walls and holes in walls, floors or ceilings around cables and other services should be fire stopped. The building itself should also continue to perform its structural functions despite fire conditions for a certain duration. Similarly, for industrial structures such as elevated platforms or equipment supports, the structure should continue to perform (as designed) even under fire conditions for a period of time. In both buildings and industrial structures, a vital area calling for passive fire protection is in the design of ***escape routes***. Passive fire protection buys time for the occupants to escape, and/or the emergency services to safely enter and stay in the building in their effort to fight the fire. Many add-on passive fire protection materials provide additional benefits like thermal and acoustic insulation.

Considering their role in enhancing life and structural safety, passive fire protection material products should be tested for their inherent fire resisting properties that include:

- Combustibility.
- Surface flame spread.
- Contribution to fire growth and spread.

Additionally, add-on passive fire protection measures are vulnerable to fire fighting activities such as the application of water jets and their condition should be inspected for their ability to perform their designed function post a fire and associated fire fighting.

An inherent concern regarding passive fire protection provisions is that they may be subjected to the 'fit and forget' mindset, except for some mechanical devices like fire doors, smoke dampers and vents, there may not be any inspection and maintenance procedures recommended either by regulatory bodies or by manufacturers. After a passive system is installed, inspected and approved by regulatory officials, the system is generally not reinspected unless structural changes affect the passive system.

Fire alarm systems can be considered part of the passive fire protection used to initiate the closing of fire doors and dampers that may be otherwise held open to allow normal building function.

6.11 SEGREGATION AND SEPARATION

As adjunct to passive fire protection measures in the built environment, various standards recommend isolation of fire hazards through separation or segregation. ***Separation*** may be interpreted as providing distance between two sections, while ***segregation*** is achieved by placing barriers between sections. ***Compartmentation*** can be explained as segregation of an area within a building such that spread of fire from within it is made difficult.

Separation distances between structures and work areas in the hydrocarbon sector are laid down in the Petroleum Rules, 2002 and OISD-STD-118 Layouts for Oil and Gas Installations. NBC 2005 and IS 1643:1988 recommend open spaces around buildings but these are not specific to fire safety. Some local by-laws of major Indian cities specify spacing of buildings primarily to address access and manoeuvrability of emergency appliances.

Segregation for fire safety is also practiced in various forms. As mentioned above, compartmentation is a form of segregation wherein designers aim to confine an outbreak of fire within an area for a period of time and prevent its spread to other areas of a building. The National Building Code 2005 in Part 4, as also IS 1641-1988 (Code of Practice for Fire Safety of Buildings (General): General Principles of Fire Grading and Classification), have classified building in terms of the activities performed in them and indicative of the nature of fire hazards as shown in Table 6.1.

Table 6.1 Typical occupancies and egress requirements recommended in urban building by-laws

Occupancy	*Group*	*Occupant load, floor area* (m^2/person)	*Occupants per unit exit width*[@]			*Max. travel distance* (m) *to exit*
			Stairways	*Ramps*	*Doors*	
Residential	A	12.5	25	50	75	22.5
Educational	B	4	25	50	75	22.5
Institutional	C	15	25	50	75	22.5
Assembly	D	0.6 to 1.5*	40	50	60	30
Business	E	3	59	60	75	45
Mercantile	F	6	50	60	75	30
Industrial	G	10	50	60	75	30
Storage	H	30	50	–	75	30
Hazardous	J	10	25	30	40	22.5

Source: Extracted from Tables 14, 15 and 16 of Bangalore Mahanagar Palike Building Bye-Laws 2003.
*Refer source code for explanations
[@] Unit exit width = 500 mm; 250 mm = additional half width

Further, NBC 2005 recommends that urban land areas be demarcated into three ***fire zones*** (1, 2 and 3 in order of their increasing fire hazard potential) with select occupancy groups as listed in Table 6.1 accommodated in them. Thus, no Group G, H and J buildings (recommended in Fire Zone 3) should be built in a Fire Zone 1 area which is primarily intended for groups A, B, C and D. Thus, segregation based on potential fire hazards is achieved. NBC 2005 also specifies four types of construction for buildings, that is, types 1–4 with decreasing stringent requirements. In relation to fire zones, the code states that buildings in Fire Zone 3 should conform to type 1 or 2 construction, Fire Zone 2 allows types 1, 2 and 3 while Fire Zone 1 can house all 4 types of construction.

6.12 COMPARTMENTATION

Within a building, segregation may be achieved through ***compartmentation,*** which involves dividing a building into parts, the bounding elements of which such as walls, floor and roof can withstand the effects of potential fire scenarios for specified durations and prevent or delay fire spread from the area first affected to other parts of the building (Figure 6.2). Any perforations or cracks on enclosing components of a compartment like walls or ceilings, deliberate (such as for air conditioning ducts, water pipes or cable passages) or otherwise, must be sealed to retain the desired fire properties of a compartment. Sealants of different types such as blocks or in situ applications like sprays are available. Fire doors of different types with appropriate fire resistance are also available. Fire doors may be manually operated or fitted with fusible links and pulleys with counterweights for automatic closing under fire conditions (Figure 6.3).

Figure 6.2 Illustration of compartmentation.

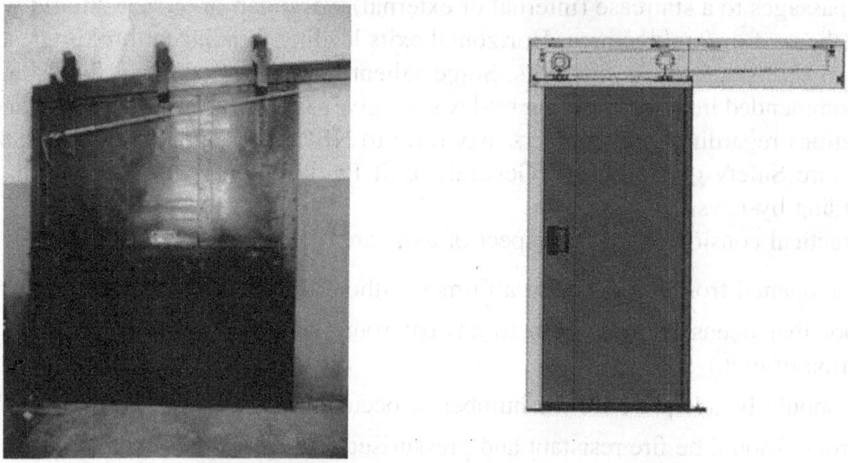

Figure 6.3 Self-closing sliding fire doors.

Compartmentation also aids in controlling the spread of fire effects and products into escape routes and refuge areas which are vital for life safety and safe evacuation of occupants.

With their well-defined boundaries, compartments have been extensively used to study different aspects of fire such as fire behaviour, heat release rate and effect of geometry on fire growth. Thus, with a vast body of knowledge about compartment fires, it is no wonder that fires in compartments was the genesis of the science of fire modelling.

6.13 FIRE ESCAPE ROUTES AND REFUGE

Safe egress of occupants from a building during a fire is a critical consideration in building design and cannot be left to be executed 'as and when' the need arises. Various codes of practice such as the National Building Code, 2005 (India), NFPA 101 Life Safety Code (USA) and Approved Document B (the UK) stipulate minimum requirements in respect of escape routes and refuge areas. The absence of dedicated and well-maintained fire escape routes in a building has caused many deaths in different types of occupancies all over the globe. The problem is enlarged many times over in places where many people could be present as in high-rise commercial buildings, large shopping malls, sports stadia, hospitals, educational institutions, etc.

An ***exit route*** should be a continuous and unobstructed path of travel from within the premises to a place of safety and comprises of three components:

- Exit access
- Exit
- Exit discharge

Exit routes must be permanent, prominently marked, separated by fire-resistant materials and with exits laid out for quick escape. They must be kept clear at all times and be available for full use in an emergency. Openings into an exit must be protected by self-closing fire doors that ordinarily remain closed or close automatically in an emergency. Exits may be doors, corridors or passages to a staircase (internal or external), verandah or terrace should with access to the street, the roof or a refuge area. Horizontal exits leading to adjacent buildings at the same level may also be considered as fire exits. Some salient features of occupant loads and exits as typically recommended in urban building by-laws are given in Table 6.1 For specific and detailed recommendations regarding exits, readers may refer to NBC 2005 and IS: 1644–1988, Code of Practice for Fire Safety of Buildings (General): Exit Requirements and Personal hazards and relevant building by-laws.

Some practical considerations in respect of exits are:

- Can be opened from the inside at all times without keys, tools or special knowledge.
- A door that opens from a room to an exit route should open outward, that is, in the direction of exit.
- Exits should be adequate for the number of occupants.
- Exit route should be fire resistant and pressurised (if required) to exclude entry of smoke or toxic gases.

6.14 PRODUCTS OF FIRE

The complete combustion of a pure hydrocarbon fuel will yield carbon dioxide and water (vapour). However, in industry or elsewhere, most combustible materials being processed, produced or used will be complex substances made up of a range of elements in widely varying concentrations with different contaminants. Further, in most occupancies, we can expect to encounter solid combustibles which may either smoulder or undergo flaming combustion.

From our earlier discussions of the fire process, we know that many solid fuels may either melt to produce vapours or undergo pyrolysis en route to vaporisation and ignition. Thus, in a fire, both decomposition products as well as products of combustion may be encountered. Sulphur or nitrogen present in combustibles can, in the process of burning, produce oxides of sulphur and nitrogen which are all harmful for people. Incomplete or inefficient combustion is more likely to generate toxic and noxious substances. Decomposition products and products of incomplete combustion include smoke, aerosols, mixtures of vapours, particulates, etc. As a rule of thumb, less efficient combustion produces more toxins, that is, substances that can cause harm to living organisms. The composite of all the products referred above is ***smoke***, comprising both visible particulates and condensates of vapours as well as invisible vapours and gases.

Some of the ‘definitions’ for smoke are:

- Particulate matter consisting of very fine solid particles and condensed vapour that constitutes the most of the visible part of the products of combustion. (Sumi and Tsuchiya)
- The ‘total effluents from a fire’ consisting of the invisible vapour phase containing narcotic fire products and the visible particulate phase that contains both irritant and narcotic fire products. (Purser)
- The airborne solid and liquid particulates and gases evolved when a material undergoes pyrolysis or combustion together with the quantity of air that is entrained or otherwise mixed into the mass. (NFPA 92B)

These different definitions agree on terms like gaseous products of combustion with liquid droplets and particulate solids and the effect of reduced visibility.

Smoke is evolved from the very early stages of a fire and smoke detectors are designed to sense this early effect. In fact, smoke detectors can sense smoke much before people can. Smoke is released from both smouldering and flaming fires but the characteristics of the smoke may vary. Smoke from flaming fires contain large particles that may coagulate to form larger particles as the fire ages; smouldering fires release smaller particles that are capable of passing through the nasal passage and entering the lungs.

Studies indicate that, for most hydrocarbons, a unit mass of fuel requires nearly fifteen times its mass of air for stoichiometric combustion and this generates sixteen mass units of smoke. But the plume from a fire will also entrain large quantities of air and thus it is difficult to actually quantify the smoke generated in a fire.

Heat is possibly the most obvious effect of fire; it can cause burn injuries on people, damage and destroy materials or help a fire to spread.

Smoke from fires can obscure visibility and cause irritation in the eyes and throat. Depending on the materials involved, fires can also release products that can harm people and property. For instance, polyvinyl chloride (PVC) insulation of cables can release hydrochloric acid, benzene, toluene and carbon monoxide, all of which are harmful to health. The type and quantity of toxic products that may be released depend on factors like the nature of the fuel, the stage of the fire, geometry of the compartment or the layout of the open, location and ventilation, while the impact on the health of both occupants and fire fighters will depend on the concentration of toxicants and duration of exposure to these hazards. Thus, a single property of a material should not be used to interpret the fire characteristics of the material.

Although laboratory conditions cannot replicate real fire conditions, scientists have, over the years, tried to assess and analyse the hazards of smoke and other associated products of fire. Designers and occupants of enclosed spaces are also similarly concerned about the fire properties of materials used in working and living areas. The field of combustion toxicology was born out of these concerns and much useful work on the subject is being done in different parts of the world.

6.15 FIRE SPREAD IN BUILDINGS

As discussed in an earlier chapter, the hot gaseous products of a fire form a plume, which tends to rise upward setting up convection currents. In the case of a compartment, the plume proceeds to become a ceiling layer, which starts growing downwards and may exit out of the compartment through any openings such as doors or ventilators. In case roof vents are installed, smoke and hot gases will exit through them. Simultaneously, colder air from outside the compartment is drawn in.

In high rise buildings, smoke and hot gaseous products from a fire rise to upper floors through any available openings and possibly cause the fire to spread to higher levels of the building. Colder air is drawn in at lower levels. If a path like a lift shaft or stairwell is available, buoyancy effects and temperature differentials between indoor and outdoor air together cause a chimney-like effect known as the stack effect and hot gases move upwards. It is for this reason that the use of lifts as fire escape is not permitted unless they are designed for such use.

The combined effect of radiation and reradiation together with induced stack effect can enhance burning rates in the case of closely placed vertical faces like high-rack storage with narrow aisles and buildings with narrow separation between them. It may not be wrong to say that while fire growth within buildings involves both convective and radiative heat transfer, fire spread from one building to another primarily involves thermal radiation.

6.16 POTENTIAL EFFECTS OF FIRE PRODUCTS ON PEOPLE

Smoke can slow down escape from a burning premises and lead to extended exposure to toxic products. Hydrogen chloride and ammonia cause direct irritation of the respiratory tract and the eyes and may serve as warning agents but, under certain circumstances, they can prevent victims from finding an exit (behavioural or physiological effects or incapacitation).

Death can be the most severe effect of fire on people but less severe and more common effects (***sub-lethal***) may include:

- Obscuration.
- Irritation and pain in the eyes.
- Strained breathing and pain in the respiratory tract. Hot smoke can damage the respiratory tract and lead to lung inflammation, bronchial constriction, respiratory distress and collapse.
- Toxic gas inhalation can cause confusion and narcosis.
- Heat from a fire can cause burns, affect the upper respiratory tract and lead to heat stroke and even collapse.

While the first two of these effects may be temporary and reversible, the last three may cause permanent damage and even death.

These effects do not occur all at once. Initially, when the fire is growing, the effect is on people's reactions and behaviour guided by factors like knowing the source of fire, apparent fire behaviour and defined role (if any). Physiological effects come next as the victim is exposed to the products of fire. Finally, death may result from exposure to heat and toxic products of fire. The potential harm from products of fire is assessed either through chemical analyses or through biological assessment. The extent of harm from a toxicant depends on factors such as its concentration in the target body organ and the duration of exposure. For narcotic fire products, its concentration in the blood supply to the brain or that inside the brain cells is critical while for irritants, the concentration in the lining of the nose, throat or lung is important.

6.17 TOXIC PROPERTIES OF FIRE PRODUCTS

In the foregoing section, we have discussed the decomposition and fire products released by different materials that may be found in different occupancies like homes and offices. Let us now take a look at the nature of hazards posed by these various products.

6.17.1 Carbon Monoxide

Most fires, particularly in areas of limited ventilation, will generate large volumes of carbon monoxide (CO) due to incomplete combustion of carbonaceous materials. CO is both odourless and colourless. When inhaled, CO combines with haemoglobin to form carboxyhaemoglobin, which replaces oxyhaemoglobin and reduces the oxygen available for the body cells. This can cause asphyxiation that may lead to death.

6.17.2 Carbon Dioxide

Carbon dioxide (CO_2) is not considered to be toxic (ACGIH TLV 5000 ppm) but can displace air resulting in an oxygen-deficient atmosphere. CO_2 inhalation increases rate of respiration and thus more of the contaminated air is inhaled.

6.17.3 Hydrogen Cyanide

Hydrogen cyanide (HCN) is produced when materials like nylon, wool, polyurethane, urea-formaldehyde and acrylonitrile-butadiene-styrene (ABS) that contain nitrogen are involved in a fire. HCN and other cyanogen compounds affect all living matter by lowering the use of oxygen by the living cells of body tissues.

6.17.4 Hydrogen Chloride

Hydrogen chloride (HCl) is produced when PVC is decomposed at fires. It forms hydrochloric acid by reacting with moisture that corrodes many materials and affects living cells.

6.17.5 Oxides of Nitrogen

There are three common oxides of nitrogen (commonly known as NO_X): nitrous oxide (N_2O), nitric oxide (NO) and the two forms of nitrogen dioxide (NO_2 and N_2O_4), the last being very toxic. Nitric oxide in atmospheric air readily converts to dioxide in the presence of atmospheric oxygen. All NO_X compounds are irritants, particularly to mucous membranes and when inhaled can damage tissues in the respiratory tract by reacting with moisture to produce nitrous and nitric acids.

6.17.6 Particulate Matter

It is reported that over 90% of particulates produced in smoke from fires are respirable and the remaining 10% either drop out of the air, or are filtered at the nose or mouth. Besides causing irritation of mucous membranes, prolonged exposure to particulate matter can affect lung capacity and lung function.

6.17.7 Formaldehyde

Formaldehyde can be smelt at very low concentrations (about 1 ppm). Exposure to low concentrations can cause irritation of eyes, nose and throat. Higher concentrations can spread irritation to the lower respiratory tract. Chronic exposure to formaldehyde may even cause cancer in the nose.

6.18 SUMMARY

Only a small number of toxicants that may be released in fires and their significant effects have been listed above. However, the ground truth is that different toxicants will be present together at a fire scenario and it is important to take note of their synergistic effects on people, assets and the environment.

QUESTIONS FOR REVISION

1. What do you understand by the term 'fire load'? What are its units? Give examples of buildings with high structural load and low fire load and vice versa.
2. What does the standard time–temperature curve represent? What are the properties tested to determine fire resistance of a building element?
3. What is the purpose of having more than one type of time–temperature curve? Discuss.
4. Name the different standards which prescribe time–temperature curves in different countries.
5. Contrast the differences in purpose of 'passive' and 'active' fire protection.
6. Discuss the means of achieving fire protection of exposed steel structures.
7. Explain the purpose of compartmentation in a building. How can the aims of compartmentation be defeated?
8. Discuss the methods of closing openings in building elements to ensure effective compartmentation.
9. What are the considerations in designing escape routes and refuge in different scenarios? Discuss.
10. Name important toxic products of building materials and contents in a fire scenario. Discuss their potential sources and their effects on people.

7

Fire Detectors and Alarms

Fire is one intruder that won't run away when the siren sounds.

—Anon

7.1 INTRODUCTION

Even when a building or facility has been designed to be resistant to ignition and fire growth, and activities are carried out with utmost care, the likelihood of a fire occurring and taking its toll cannot be totally eliminated. To protect life and property in case of fire, any premises should have means of alerting occupants about an outbreak of fire at an early stage and thereafter to control the spread of fire and extinguish it. Thus, in addition to passive fire protection provisions such as facility design and selection of materials of construction, there must be appropriate means of alerting people, controlling fire spread and possibly extinguishing it in case a fire breaks out. The various devices and systems that may be installed could be based on regulatory provisions, engineering guidelines, perceived threats and resources available (Figure 7.1). It may be impractical and awfully expensive to install fire protection facilities to fight catastrophic fires; a good system may be the one that aims to prevent emergencies from developing into major threats to life and property within a facility and in its surroundings. The different types of active fire protection in any occupancy may include one or more of the following:

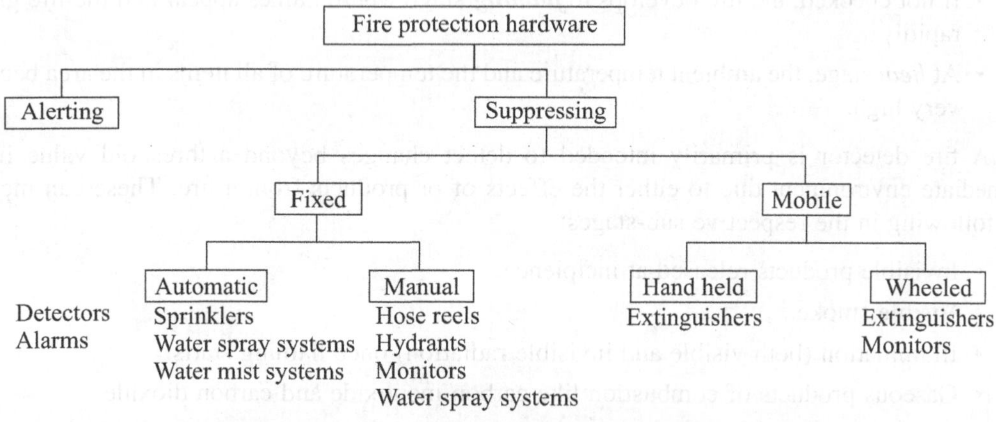

Figure 7.1 Types of fire protection hardware.

- Fire detection and alarm systems
- First-aid fire fighting equipment
- Water-based fixed fire protection systems
- Fixed systems based on media like carbon dioxide and vaporising liquids
- Mobile fire fighting appliances
- Communication systems

In this chapter, we will confine our discussions to the construction and working principles of different types of fire detectors. Detectors and alarms may either only alert occupants or emergency services to the occurrence of fire conditions or they may be designed to actuate some fire control system.

7.2 BASICS OF DETECTORS AND ALARMS

When people are around and alert, human senses can detect some effects of fire such as heat, smoke or light. People may also smell the burning of different materials or hear crackling sounds. However, in places where people are not present or when they are not alert as while sleeping or are not mobile due to health conditions, fire detectors are useful devices. They act by sensing one or more of the effects (thermal as well as non-thermal) or products of fire and raise an alarm either locally or on a panel located elsewhere. Detectors are classified based either on the fire effects they sense or on their working principles.

In an earlier chapter, we discussed the stages of the fire process viz., pre-ignition, ignition, growth, steady burning and decay. The growth stage itself may be divided into the following sub-stages:

- ***Incipient*** when the fire has just started and is confined to the material first ignited. Products of combustion are invisible at this stage.
- ***Smouldering*** when smoke appears and the rate of burn becomes noticeable (This stage may not take place in liquids or solids that melt and then burn like liquids).
- If not checked, the fire develops to ***flaming*** stage where flames appear and the fire grows rapidly.
- At *heat* stage, the ambient temperature and the temperature of all items in the area become very high.

A fire detector is primarily intended to detect changes beyond a threshold value in its immediate environment due to either the effects of or products from a fire. These can include the following in the respective sub-stages:

- Invisible products released at incipience.
- Visible smoke.
- Illumination (both visible and invisible radiation) once flaming starts.
- Gaseous products of combustion like carbon monoxide and carbon dioxide.
- Rise of ambient temperature to a set point or its rapid rate of rise.

While it is important that a fire should be detected as early as feasible for life safety and damage control, it is worth remembering that it is not easy to detect very early conditions of incipience.

The passage of development through the sub-stages of fire growth is determined largely by the fuel, its configuration, ambient conditions and the nature of burning. These stages occur almost instantaneously in an explosion, while a smouldering fire can smoulder for days and may or may not develop into a full-blown fire. All of them, however, follow the same route and fires may be detected at any stage en route.

7.3 DETECTOR TYPES BASED ON EFFECTS

Detectors may also be classified as 'spot' and 'line' types. Spot detectors have the sensor located at one point and can sense the actuating condition within a limited radial area around it. Examples of spot detectors are bimetallic detectors, fusible alloy detectors, some pneumatic rate-of-rise detectors, some smoke detectors and thermoelectric detectors. Line detectors may be located either at one point or over a continuous distance but can, in both cases, sense the actuating conditions over the designed distance. Examples of line type detectors are rate-of-rise pneumatic tubing detectors, projected beam smoke detectors and heat-sensitive cable. A broad classification of detectors is given in Table 7.1 and their working principles and sensitivity are shown in Figure 7.2.

Table 7.1 Broad classification of sensors

Heat	Fixed temperature (expansion, melting/fusion, resistance thermistor)
	Rate of rise of temperature
	Combination
	Rate compensation
Smoke	Ionisation
	Photoelectric
	Aspirating
Gases	Carbon monoxide
Light	Visible light – light obscuration
	Visible light – light scattering
	IR radiation
	UV radiation
Gas	Carbon monoxide

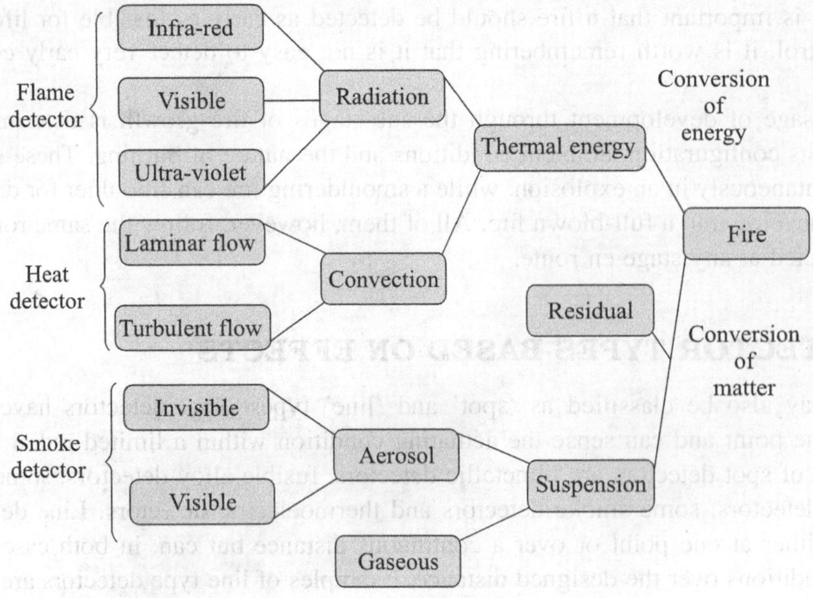

Figure 7.2 Types of fire detectors based on detection principles and sensitivity.

The area that may be covered by spot detectors will depend on the type of detectors, compartment configurations, ambient conditions, etc. The same detector when installed in a large open room can provide different coverage than when installed in a corridor or passage. In very large spaces like atria, aircraft hangars and indoor stadia, line detectors are the preferred type and the area covered will also depend on numerous factors.

7.4 HEAT DETECTORS

Fires are known to have two types of thermal effects:

- Rise of temperature in the immediate surroundings, and
- Faster rise of ambient temperature than that due to normal atmospheric changes.

Both of these effects are utilised as principles of actuation of heat detectors.

7.4.1 Fixed-Temperature Heat Detector

A fixed-temperature heat detector is actuated when its sensor detects a set temperature. We know from discussions in earlier chapters that the products of combustion ascend to higher levels by convection. Thus, regions above the fire tend to have high concentrations of fire products besides being hot. In a location that is confined at a height as by a ceiling in a compartment, the effects of convection will result in the formation of a hot ceiling layer that grows in thickness with time. The greater the dimensions of the ceiling, the longer it will take for the hot ceiling layer to form for the same size of fire and then exceed threshold values to actuate sensors of detectors.

Fixed temperature type detectors are the most common heat detectors in use and are set to operate when the temperature around the detector reaches a set value that is well above the maximum temperature reached at that location under normal conditions. Some sensing elements employed in fixed temperature detectors are

- Fusible elements made of alloys or eutectic metals that melt at a known temperature and soldered joint type sprinklers.
- Expanding metal or a gas contained in a closed space that complete an electric circuit.
- Quartzoid bulb type sprinklers where a liquid expands to shatter the bulb and release water held in the system.
- Bimetallic elements where differential expansion of the two metals causes the element to bend and make (or break) an electric circuit.
- Line detectors made up of two cables with an insulated sheathing that breaks down when heated to complete an alarm circuit.

The detecting element in a fixed temperature detector must be fully heated to its set temperature for the alarm to be actuated and this makes a fixed-temperature detector suited for slow-growing fires. For fast-growing fires, the detector element may not be heated fully as is required for fixed temperature detectors and a dangerous time lag may occur. Fixed-temperature detectors are suited for spaces that may normally have high ambient temperatures such as in kitchens, garages or bathrooms.

Fusible Element

Some alloys and eutectic metals (e.g., bismuth, lead, tin and cadmium) melt at relatively low temperatures (generally in the range 55–180°C), a property that has been used in one type of fixed-temperature detector. The fusible element can be a solder to hold a spring under tension. When the element fuses, the spring is released to complete a circuit and initiate an alarm. Soldered-joint type sprinklers and automatic fire door closers use fixed temperature detector based on fusible elements (Figure 7.3). It is obvious that fusible link type detectors are not reusable.

Figure 7.3 Typical fusible link sprinkler and door closer.

Bimetallic Elements

When a bimetallic strip made of two metals with different coefficients of thermal expansion (e.g., iron and copper) is heated, one expands more than the other. This differential expansion causes the strip to bend and the effect can be used to close a normally open circuit or open a normally closed circuit (Figure 7.4). The size of the gap between the contacts determines the temperature to which the strip must be heated to get the desired effect. Detectors using bimetallic strips generally comprise invar (a Ni-Fe alloy) for the low-expansion component and alloys of Mn-Cu-Ni, Ni-Cr-Fe, or stainless steel for the high-expansion component.

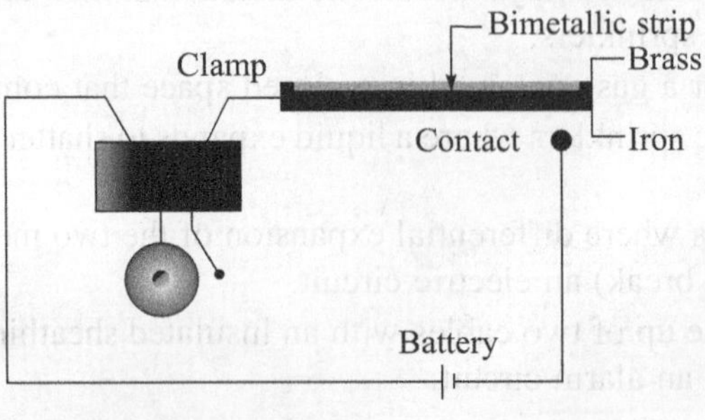

Figure 7.4 Schematic of bimetallic strip fixed-temperature type detector.

A big plus of heat detectors based on thermal expansion of metals is that they are self-restoring after operation, when the ambient temperature drops below the set point.

Break-line Cable

A length of flexible PVC insulated cable is cut at intervals and the bared conductors are rejoined with a low-melting fusible alloy. Under normal conditions, there is a steady flow of current through the conductor but during a fire, the fusible joint melts and this break in circuit actuates an alarm. This is a simple form of a line detector.

Expansion of Gases

Gases have a high coefficient of expansion and this property has been used in pneumatic heat detectors. A known gas (commonly air) is filled into a chamber with a flexible diaphragm. A rise of ambient air temperature expands the air in the chamber which exerts pressure on the diaphragm. Sustained expansion pushes up the diaphragm until it completes an electrical circuit and raises the alarm (Figure 7.5).

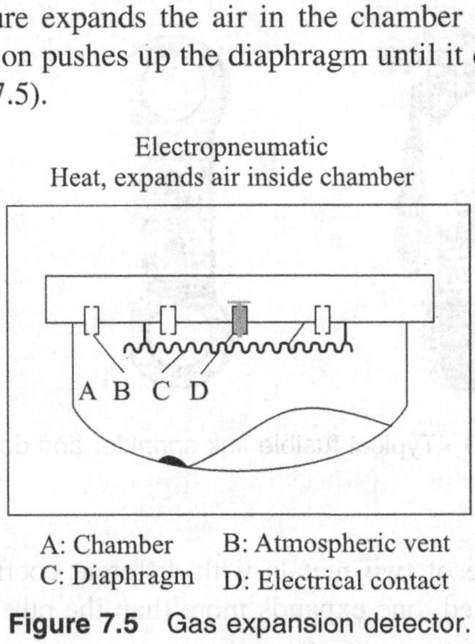

Figure 7.5 Gas expansion detector.

Expansion of Liquids

The quartzoid bulb sprinkler is an established example of heat detector based on thermal expansion of liquids. In such sprinklers, both the liquid and the air bubble contained within the liquid expand when there is a rise in temperature of the surrounding air. Under sustained fire conditions, the glass bulb shatters and water is sprayed over the design area (Figure 7.6).

Figure 7.6 Frangible bulb sprinkler.

7.4.2 Rate-of-Rise of Temperature Heat Detector

Fast-growing fires lead to a rapid rise of temperature in the immediate surroundings which a rate-of-rise (ROR) heat detector is designed to sense. The internal components of such a detector constantly compare the temperature of the surroundings to a baseline temperature programmed into the detector and the detectors are designed to compensate for any normal variations in ambient temperature. A set of predetermined criteria are also programmed into the detector and when the rise in temperature matches these criteria, the alarm is actuated (Figure 7.7). A typical ROR heat detector responds when temperature rise is in the range 7–8°C/min or higher.

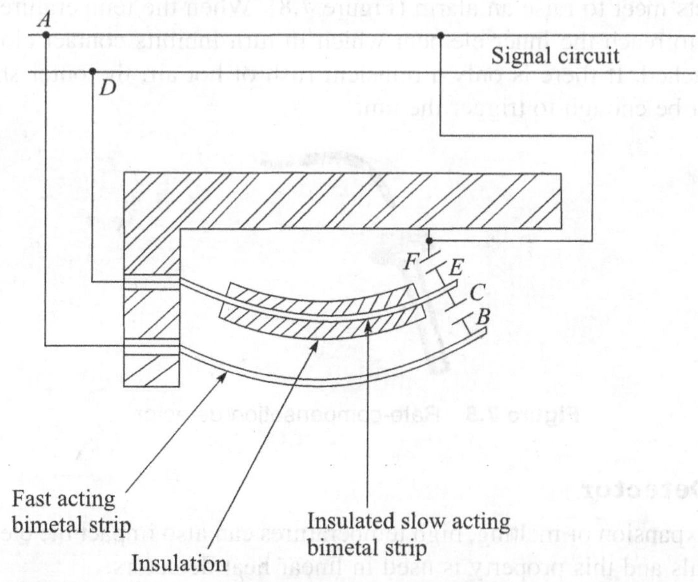

Figure 7.7 Schematic of rate-of-rise of temperature detector.

An ROR heat detector is best suited for areas that are inaccessible to smoke or where the average temperature remains constant but are likely to give false alarms in case of events like rush of hot air when oven doors are opened.

7.4.3 Combination Heat Detector

To take advantage of the positive features of both, fixed-temperature and ROR detectors, designers have developed a combination-type heat detector. The rate-of-rise element responds to rapidly growing fires, while the fixed-temperature component reacts to slow-burning fires. If the air chamber (of the gas expansion detector discussed above) has a fine, accurately designed vent, some expanding air will be vented out to compensate for any temperature rise such as normal increases in ambient temperature. When there is a rapid rise in temperature, the air in the chamber expands faster than the compensating vent can release it and the expanding air pushes the diaphragm against the electrical contact to complete the circuit and raise the alarm.

A combination heat detector offers good protection for areas with unstable, high average temperatures, such as rooms housing several ovens that are routinely opened and closed.

7.4.4 Rate Compensation Detector

A rate-compensation detector has a big advantage over both fixed temperature and ROR types of detectors because it can accurately sense the surrounding air temperature regardless of the fire growth rate and the system is activated at the set temperature.

A typical rate-compensation detector comprises a pair of low thermal expansion metal or alloy struts housed in a tubular outer shell of a high expansion metal or alloy, which closely follows changes in surrounding air temperature. The struts have a contact on each which stay apart under normal conditions. When exposed to a rapidly growing fire, the outer shell tends to expand, while the inner struts are slow to expand and exerts restraint on the outer shell. In due course, the contacts meet to raise an alarm (Figure 7.8). When the temperature rise is slow, the heat takes longer to reach the inner element which in turn inhibits contact closure till its rated temperature is reached. If there is only a transient rush of hot air, the outer shell may expand, but that would not be enough to trigger the unit.

Figure 7.8 Rate-compensation detector.

7.4.5 Line Detector

Besides causing expansion or melting, high temperatures can also impact the electrical properties of various materials and this property is used in linear heat detectors.

The core of a linear heat detector is a pair of twisted wires of different metals or alloys to give them the right strength, electrical conductivity and corrosion resistance. Each wire is insulated with material that loses its insulating property at higher temperatures. The wire pairs have an outer jacket to protect against damaging environmental conditions (Figure 7.9). When the rated temperature is reached at any point along the line detector, the wires contact each other to create a short circuit and actuate an alarm.

Figure 7.9 Line detector.

7.4.6 Heat Detector – Summary

Heat detectors are reliable, inexpensive and easy to maintain. However, a limitation of fixed-temperature type detectors is that they are activated only when room temperatures are fairly high and, by then, the fire could be well-developed with rapidly growing damage. Hence, fixed temperature detectors may not be the right choice for locations where early detection is desired such as in locations where life safety is a primary concern or where high-value heat sensitive contents are housed. As for heat detectors based on fusible links or sprinkler type devices, they are single-use type where a part or whole of the device may need to be replaced after it has actuated.

7.5 SMOKE DETECTORS

In an earlier chapter, we have described the generation and composition of smoke in the fire process. Vapours are produced even prior to ignition during the process of pyrolysis of solid fuels or vaporisation of liquid fuels. At incipience, most fires produce smoke, but the character of the smoke varies with fuel and prevailing conditions. Smoke detectors are intended to respond to a threshold concentration of 'smoke', which includes pyrolysis products, vapours, visible or invisible products of combustion, etc. Two common types of smoke detectors are the ***ionisation*** and ***photoelectric*** types. Both may appear the same on the outside but their construction and working principles are different.

7.5.1 Ionisation Type

Ionisation type smoke detectors contain a small amount of radioactive material (commonly a foil of americium oxide in gold matrix) in a silver capsule held inside a palladium chamber. Americium-241 is a good source of alpha particles. The chamber wall is thick enough to retain the radioactive material, but thin enough to allow the passage of alpha particles.

The ionisation chamber contains two metal plates separated by a small distance and connected to a battery. One of the plates carries a positive charge (anode) and the other, a negative charge (cathode). Air molecules between the two plates get ionised as electrons are kicked out of the molecules by alpha particles from the radioactive material. Oxygen and nitrogen (of the air) ions are positively charged being short of electrons and are attracted to the cathode; the free electrons are negatively charged and are attracted to the anode. When smoke enters the ionisation chamber, it disrupts this current. The detector senses the drop in current between the plates and sets off the alarm [Figures 7.10(a) and (b)].

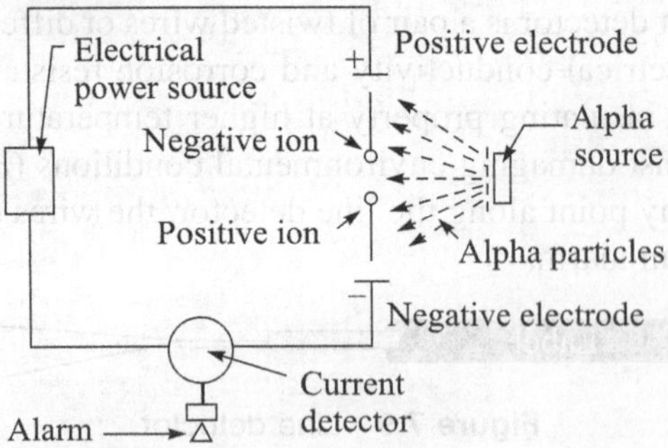

Figure 7.10(a) Schematic of ionisation type smoke detector–working principles.

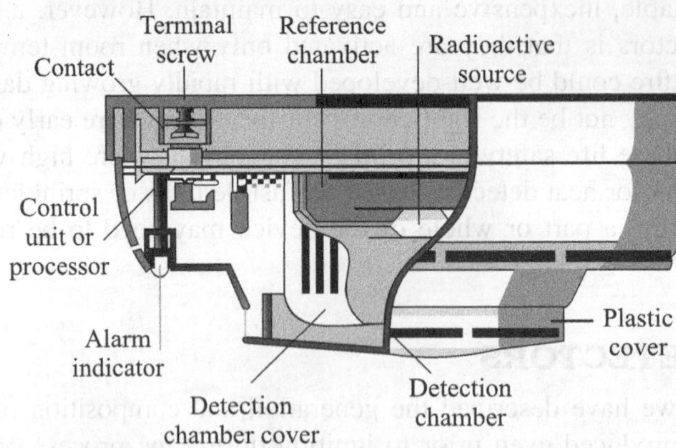

Figure 7.10(b) Typical ionisation type smoke detector–construction.

In dual chamber type detectors, there are two chambers one open to the outside and the other closed. In normal conditions, the current flowing through both chambers is the same. Under fire conditions, smoke particles enter only the open chamber and there is a resultant imbalance in the current flowing through the two chambers and this imbalance is sensed in this type of ionisation smoke detectors.

Ionisation type smoke detectors are usually of the spot type. They are very sensitive in the early stages of a fire when smoke particles are small and hence must be sited very carefully. In some locations such as a garage or kitchen, the products of combustion could be present in 'non-fire' conditions and installing ionisation detectors in those areas could give repeated false alarms. It is also important that these detectors are not placed near a ventilating or fresh air inlet where a current of clean air can pass over them and inhibit their speed of reaction in a fire situation.

7.5.2 Photoelectric Type

As the name implies, photoelectric type detectors work on the principles of the amount of light reaching a photoelectric cell. The two sub-classes of photoelectric type detectors are briefly described below.

Light Obscuration Type

In ***light obscuration*** type detectors, light from a source such as an LED is normally incident on a photosensitive receiver (photodiode, phototransistor, etc.). When smoke enters the path of light between the source and the receiver, less light reaches the receiver than normally and, when the smoke density/obscuration reaches a set level, an alarm is actuated (Figure 7.11). Light obscuration smoke detectors are generally beam type and are installed with the light source at one end of the area to be protected and the receiver at the other end. They are used to protect large open areas.

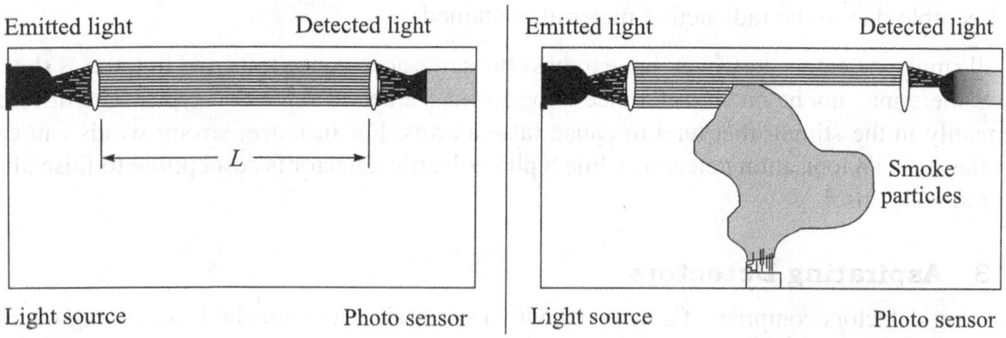

Figure 7.11 Principle of smoke detection by light obscuration.

Light Scattering

Light scattering type detectors are built such that under normal non-fire conditions when the detector chamber is free from any smoke particles, light emitted from the light source does not reach the sensor (photocell). When smoke enters the chamber, some of the light is scattered toward the sensor and, when the smoke density reaches a set threshold, the light reaching the cell will either create an electrical current in the detector circuit or allow more current to flow through it and an alarm will sound (Figure 7.12). Smoke detectors utilising the photoelectric light scattering principle are usually of the spot type.

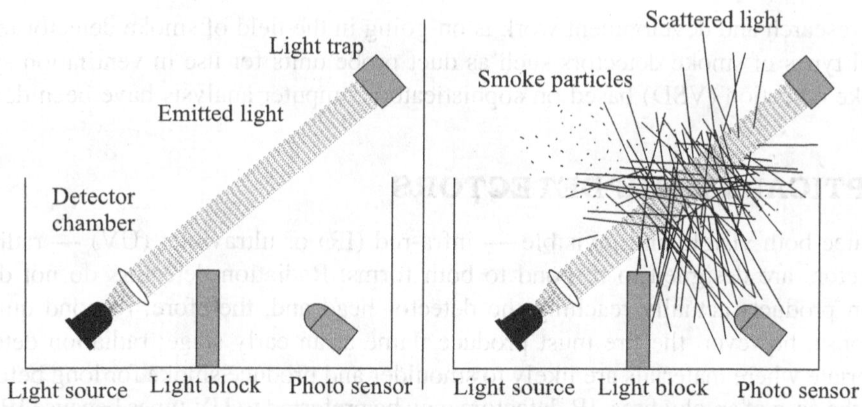

Figure 7.12 Principle of detection of smoke by scattered light.

Comparing Ionisation and Photoelectric Detectors

Both types of smoke detectors have their own plus and minus points as given below:

- Ionisation detectors can sense a wider range of smoke than photoelectric detectors.
- Photoelectric detectors are more sensitive to smoky fires with larger smoke particles, while ionisation detectors are good at detecting flaming fires with smaller smoke particles.
- Dark smoke has poor reflecting properties than white smoke and this can affect the performance of light scattering type photoelectric smoke detectors.
- Ionisation smoke detectors are relatively inexpensive but disposal of old detectors is not simple due to the radioactive material contained.

All smoke detectors, however, have to pass the same acceptance tests and in today's state-of-the-art, there may not be much difference in performance between the two types. The difference lies mainly in the stimuli that tend to cause false alarms. For instance, strong winds can cause false alarms in an ionisation detector, while a photoelectric detector is susceptible to false alarms from radio signals.

7.5.3 Aspirating Detectors

Aspirating detectors comprise of a detector with a small air pump attached to a piping manifold with perforations used to draw air samples from the surroundings and send it to the sensing element. They can be described as a combination of the point and line-type smoke detectors. The air being sampled may be passed through a filter before being analysed for the presence of smoke. Very sensitive sensors are selected for use in aspirating detectors to account for the effects of dilution of smoke with air as it is aspirated.

7.5.4 Duct Detector

Another similar type of detector is the 'duct detector' used to monitor the quality of air as it travels through building HVAC ductwork. It has the advantage of using a single detector to sample a much greater volume of air than would be possible with an ordinary spot type smoke detector.

Much research and development work is on-going in the field of smoke detector technology and special types of smoke detectors such as duct probe units for use in ventilation systems or video smoke detection (VSD) based on sophisticated computer analysis have been developed.

7.6 OPTICAL FLAME DETECTORS

Fires produce both visible and invisible — infra-red (IR) or ultraviolet (UV) — radiation and flame detectors are designed to respond to both forms. Radiation detectors do not depend on combustion products actually reaching the detector head and, therefore, respond quickly. For early response, however, the fire must produce flame at an early stage; radiation detectors are not appropriate where materials are likely to smoulder and produce smoke for long before flames appear. In the case of smoky fires, IR detectors may be preferred to UV types because IR radiation can penetrate smoke better.

Infrared (IR) detectors primarily comprise a filter to filter out unwanted wavelengths and a lens to focus the incoming energy on a photocell sensitive to IR radiation. A negative of IR detectors is that solar radiation can cause interference and some shielding may be required where solar radiation may be present. Ultraviolet (UV) detectors are relatively free from the effects both of sunlight or artificial light.

Flame detectors can be used to protect large open areas. They act fast but require direct line of sight to all parts of the area covered. Flame detectors respond rapidly to fires in clean burning fuels such as alcohol or methane that would not be detected by smoke detectors.

At places where flames may take long to appear, it may be useful to install flame detectors in conjunction with smoke detectors. Then, one type of detector may respond before the other in case of mixed contents within a compartment. As to whether the alarm or fire control systems should be actuated when only one detector type or both have sensed fire effects is a decision the premises management has to take.

7.7 GAS-SENSING DETECTORS

Studies show that, at a fire scene, detectable quantities of some gases are released after smoke and before heat can be detected. Fire gas detectors use either semiconductor or catalytic element technology. Both oxidising and reducing gases create electrical changes in the semiconductor. In the latter, the contained catalyst accelerates the oxidation of combustibles and raises the temperature of the element to actuate an alarm.

Carbon monoxide (CO) detectors are sensitive to the gas given off in the early smouldering stages of a fire. They are not affected by ambient dust and steam (which can result in false alarms in smoke detectors) and are recommended for sleeping areas.

7.8 APPLICATION-SPECIFIC DETECTORS

Conventional smoke detectors are go/no-go devices implying that no information other than alarm or no-alarm is transmitted to the alarm system. In such detectors, therefore, it is essential that the sensitivity of the detector match the expected conditions of the location. The use of microprocessors in fire alarm systems has helped overcome some such limitations. For instance, sensors using analogue technology are used to monitor conditions in a protected area and transmit it to a computer-based alarm control unit. This permits system designers to adjust sensor sensitivity to suit ambient conditions where necessary. Analogue sensors can be photoelectric or ionisation, or combination of thermal, photoelectric and ionisation units.

It has been a practice in fire detection system designed to 'cross zone' different types of detector referring to installing a combination of different types of detectors to address the limitations of each type and is practiced more where an automatic release of extinguishing agents is built in. This led to the development of multisensor multicriteria capability type detectors also known as *application-specific detectors* (ASD). The purpose of combining sensors in this way is to enhance the detection performance or its resistance to certain types of phenomena likely to cause a false alarm or both. Application-specific detectors have found application in special areas like computer rooms, atria, HVAC ducts (where high air velocities are prevalent), welding shops, parking garages, high-bay warehouses, etc.

7.9 SELECTION OF DETECTORS

The final choice of detector type has to be based on the risk to be protected and the individual circumstances of each case. Many factors such as the size of the room, height of the ceiling, composition of combustible materials, activities that may cause false alarms (such as using an arc welder in a room protected with a UV or visible flame detector) and the air handling system (HVAC) have to be considered in the selection of detectors. Whatever detector is employed, however, it needs to be reliable, robust and economical.

7.10 ALARM SYSTEMS

The simplest and oldest fire alarm probably is the fire bell. Public address (PA) systems are also widely used in industries and large buildings to alert occupants about an outbreak of fire and to advise them about evacuation or similar actions. The century old sprinkler system has an inbuilt bell that is set ringing by the flow of water. The break glass fire alarm is a common manually actuated device that may have complex electronics for annunciation (Figure 7.13). Stand-alone single station detectors like those used in homes have an in-built alarm. In large multioccupancy or high-rise premises where occupants in one section may not otherwise be aware of an outbreak of fire in another section, a detector actuated in one section may require to be intimated to people in other areas. Today's alarm circuitry and panels use complex electronics and larger systems require linkages for power supply, alarm systems, automatic shut-off devices, fire control systems and possibly even between detectors.

Figure 7.13 Break glass fire alarm call point.

A fire detection and alarm system may include some or all of the following:

- Alarm initiators, e.g, automatic fire detectors, manual pull stations, break glass alarms, sprinkler system flow devices, etc., connected to the system control unit.
- Alarm indicators (audible or visual) connected to the system control unit.

- Auxiliary controls such as ventilation shut-down, smoke control or to activate a fire protection system connected to output circuits of the system control unit.
- System control unit.
- Primary electrical power supply as well as a stand-by power supply (batteries or emergency generator).
- Signal to a central 24 × 7 manned station or an external response location, like a fire station.

7.11 FIRE ALARM SYSTEMS AND CONTROL PANEL

Francis Upton patented the first electric fire alarm in 1890 in USA but major improvements came only seven decades later in 1960s. A fire alarm control panel is a core component of a fire alarm system and can rightly be called the 'CPU' of a fire detection and alarm system (Figure 7.14). The primary purpose of an annunciator panel is to monitor each circuit, zone or detector for condition of alarm or other abnormality, display the status of that condition and to operate any required output or outputs as per system design and are intended to warn occupants of a fire signal and notify emergency response personnel at security or fire station. It may also be linked to the public fire brigade station or to fire control systems.

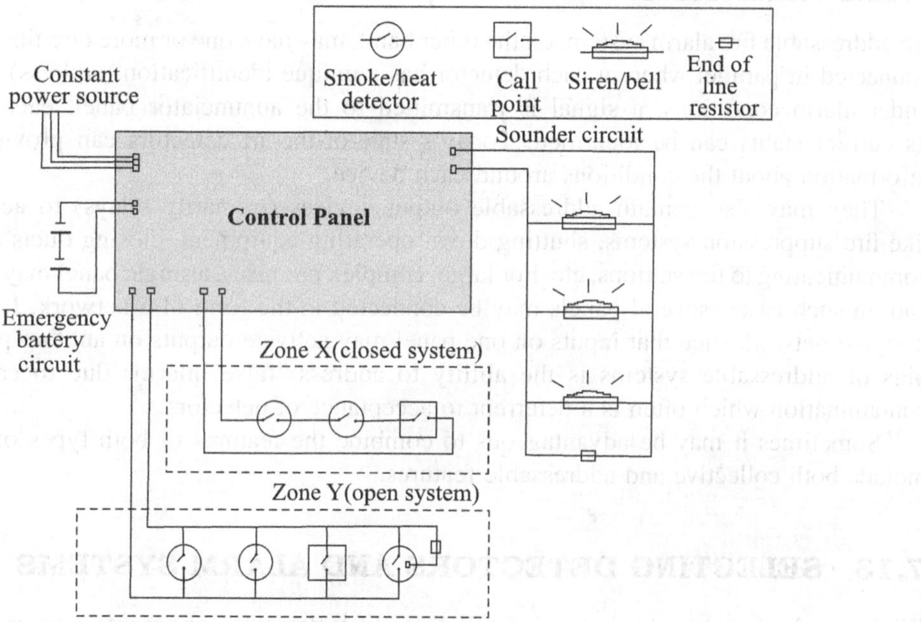

Figure 7.14 Typical fire alarm system components.

Generally it may be located at a monitored location like the security post of a building or at an industrial fire station. It may also be connected to a remote location like a public fire station through various means of communication. Control panels may be described as conventional panels and analogue addressable panels.

Two types of panels in use today are:

- Conventional or collective and
- Intelligent or addressable

Detectors in a protected area are, for practical reasons of annunciation on panels, divided into zones. A zone may contain either one or more detectors grouped by location or all the detectors in a collective circuit on that circuit. Standards limit the number of detectors in a zone.

7.12 PRINCIPLE OF OPERATION

7.12.1 Collective

A collective fire alarm system comprises one or more circuits, with detectors connected in parallel. An alarm signal is received at the fire panel as the current on a circuit changes beyond set limits when one of the detectors on that circuit activates an alarm. The control panel of a collective system can display the alarm or fault status of a circuit but cannot identify the status of individual detectors of the circuit.

7.12.2 Addressable

An addressable fire alarm system, on the other hand, may have one or more circuits with detectors connected in parallel wherein each detector has a unique identification (address). Hence, when under alarm conditions, a signal is transmitted to the annunciator panel, each detector and its current status can be identified. Today's state-of-the-art detectors can provide a range of information about the conditions around each device.

They may also contain addressable output devices (primarily relays) to actuate services like fire suppression systems, shutting down operating equipment, closing ducts and openings, communicating to fire stations, etc. For large, complex premises, a single panel may be inadequate and, in such cases, several panels may be connected in the form of a network. It is possible to design a network such that inputs on one panel may activate outputs on another panel. Another plus of addressable systems is the ability to address 'false' alarms due to causes such as contamination which often is a deterrent to acceptance of detectors.

Sometimes it may be advantageous to combine the features of both types of systems that include both collective and addressable features.

7.13 SELECTING DETECTORS AND ALARM SYSTEMS

While deciding on the selection and installation of fire detectors, occupants, owners and managers should understand both smoke detection systems and know the alternatives available. A rule of thumb is that for life safety, early detection is most important and hence, smoke detectors are preferred. Heat detectors are activated in the later stages of a fire and may be appropriate when property protection is a major concern. As mentioned above, cross zoning may also be adopted particularly in systems designed for automatic discharge of extinguishing agents.

It is important to note that both optical flame and heat detectors cannot detect smoke. They can only react to a flaming fire, not an incipient one.

A very special consideration for selecting detectors is when one or more hearing impaired persons could be sleeping.

QUESTIONS FOR REVISION

1. Would you categorise detectors as active or as passive means of fire protection? Justify your answer.
2. Detectors may be classified on the basis of fire effects they detect. Explain.
3. Explain the term 'cross zoning' of detectors. What could be a situation where you would recommend cross zoning?
4. Describe the working principles of different types of detectors.
5. On what bases would you select the type(s) of detectors for a given occupancy?

8

Fire Extinguishers

Putting fancy tools into poorly-trained hands simply makes for more expensive errors.

—Firehouse Fire Department quotes

8.1 FIRE FIGHTING FUNDAMENTALS

Once a fire breaks out, efforts must be made to control its growth and possibly extinguish it. It may be possible to put out a very small fire on materials like grass or paper either by stamping on it or by beating it with a fire beater but there are better ways of putting fires out like using fire buckets or hand-held extinguishers. For larger fires or to cover a wider area, there are hydrants, sprinklers and other fixed systems or mobile appliances like fire tenders of different types.

NFPA 750 has the following explanations related to fire fighting:

- ***Fire control*** refers to limiting fire growth by pre-wetting adjacent combustibles and controlling ceiling gas temperatures to prevent structural damage.
- ***Fire suppression*** is the sharp reduction of heat release rate of a fire and prevention of regrowth.
- ***Fire extinguishment*** implies complete suppression of a fire until no combustible materials are burning.

8.2 FIRST-AID FIRE FIGHTING EXTINGUISHERS

As the name implies, first-aid fire extinguishers are meant for use by occupants to fight fires in their early stages even before the arrival of professional fire fighters. Hence, it is essential that trained personnel are available on the premises at all times for fire extinguishers to be of practical utility in an emergency. Fire extinguishers perform by removing or inhibiting one or more of the four faces of the fire tetrahedron. Their selection, construction and working principles as given in BIS standards are explained below.

IS:2190-2010, Code of practice for selection, installation and maintenance of first-aid fire extinguishers is the basic Indian guidance document on the subject of hand-held fire extinguishers. Water, carbon dioxide and mechanical foam type fire extinguishers of capacity larger than 50 litres are called *mobile fire extinguishers* and, therefore, do not come under the purview of this code of practice. However, the standard applies for mobile extinguishers also so far as classification of fire for selection of extinguishers is concerned.

The code describes the suitability of different types of extinguishers based on the material that is burning and are classified accordingly. It also includes guidelines for inspection, testing, maintenance, recording and life of extinguishers. Different extinguishing media are suited for the specific mode of extinguishing required for different fuels. All fire extinguishers (except carbon dioxide) contain the extinguishing agent specific to its type that is expelled by pressure either due to pressurised gas/air contained within the extinguisher shell or due to release of gas from a cartridge contained within it. Larger extinguishers may have the cartridge mounted outside the extinguisher shell. Carbon dioxide extinguishers basically are gas cylinders with provision to discharge the gas through a horn made of non-metallic material. Each of the extinguisher types also has its respective specification of construction and performance and larger extinguishers are mounted on trolleys for easy manoeuvrability.

Another significant Indian document on fire protection is the Fire Protection Manual of the Tariff Advisory Committee which is applicable for assets insured against fire.

8.3 COMMON FEATURES OF EXTINGUISHERS

Portable extinguishers contain extinguishing media that require a moving force to be expelled. In the case of water, foam or dry chemical powder (DCP) extinguishers, this force is derived either from pressurised air above the contents (stored pressure type) or from a cartridge containing carbon dioxide (CO_2) gas (gas cartridge type). Metal dry powder and halocarbon extinguishers should be charged only with the type of gas mentioned in the label of the extinguisher. CO_2 extinguishers are simply CO_2 cylinders with an appropriate discharge horn. The size of the cartridge depends on the size of the extinguisher and may accordingly be located inside the extinguisher or attached outside it.

Stored pressure type extinguishers are filled with dry air or nitrogen (both of specific quality) till the desired pressure is built-up in the shell and the container is sealed tight. A pressure gauge is mounted at the top of the extinguisher shell to indicate availability of adequate pressure for use. While actuating, the safety pin is withdrawn and valve lever is depressed to force out the extinguishing medium through an internal siphon tube into the delivery hose and nozzle (Figure 8.1).

Gas cartridge type extinguishers have a CO_2 cartridge with a sealing disc screwed to the cap of the extinguisher. The plunger has a piercing nail attached to it which punctures the disc of the cartridge when pushed down. CO_2 gas pressurizes the extinguisher up to 6–7 kg/ cm^2 pressure and expels the contents of the extinguisher as in the case of stored pressure type extinguishers. The cartridge must be weighed periodically as per standards and changed if mass loss is more than 10% of the original. The cartridge must also be replaced after an extinguisher is used.

Different types of extinguishers may look similar externally but they are different and are manufactured to meet their respective specifications (Table 8.1). Hence, filling an extinguisher meant for one medium with another must never be attempted. Similarly, powders of different types used in DCP extinguishers must never be mixed or interchanged. Both types of extinguishers described above are meant to hold high pressures and hence need to be designed and properly maintained.

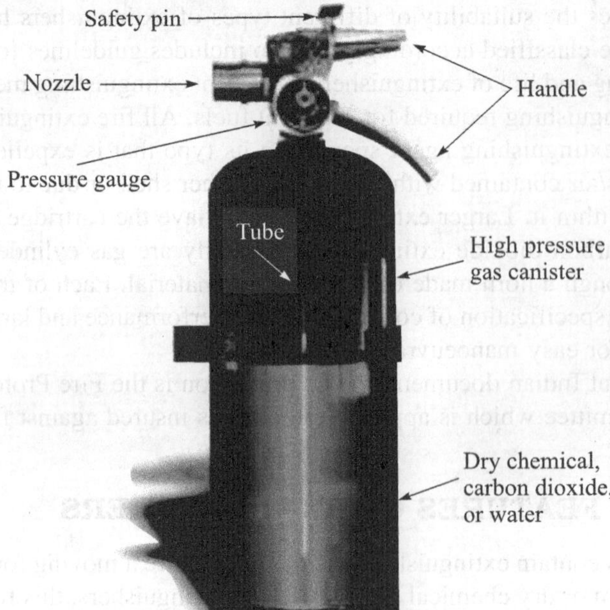

Figure 8.1 Typical components of an extinguisher

Table 8.1 Extinguisher — methods of expelling

Extinguishing materials	*Expelling methods*		
	Self-expelling	*Gas cartridge or cylinder*	*Stored pressure*
Water			✓
Foam		✓	✓
Carbon dioxide	✓		
Dry chemical powder		✓	✓
Clean agent	✓		

8.4 TYPES OF EXTINGUISHERS

There are different types of extinguishers which are described below:

8.4.1 Water Extinguishers

Fires on ordinary combustible materials are best extinguished by cooling action which is commonly effected by water. Accordingly, extinguishers that discharge water are commonly installed in different occupancies where Class A type fires are more likely. The extinguisher shell contains water with some additives (Figure 8.2). The reach of the water stream is about 6 m and a 9 litre discharge should last for at least two minutes.

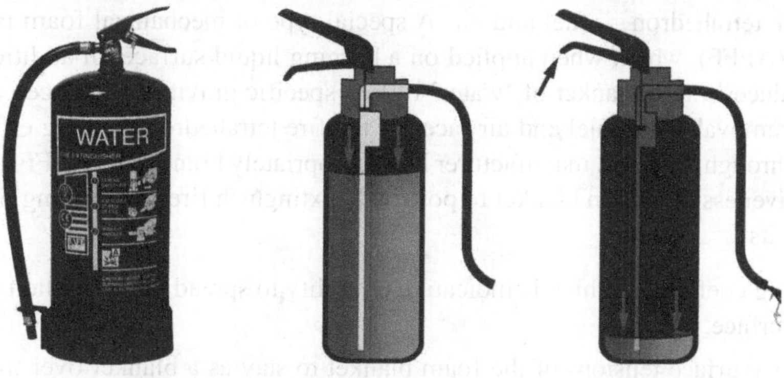

Figure 8.2 Water extinguisher — external and sectional views.

8.4.2 Foam Extinguishers

The construction of a foam type extinguishers is very similar to that of water-type extinguishers except the discharge nozzle has a small foam-making branch. As for contents, instead of water, a foam extinguisher contains foam compound mixed in water in the required ratio (Figure 8.3).

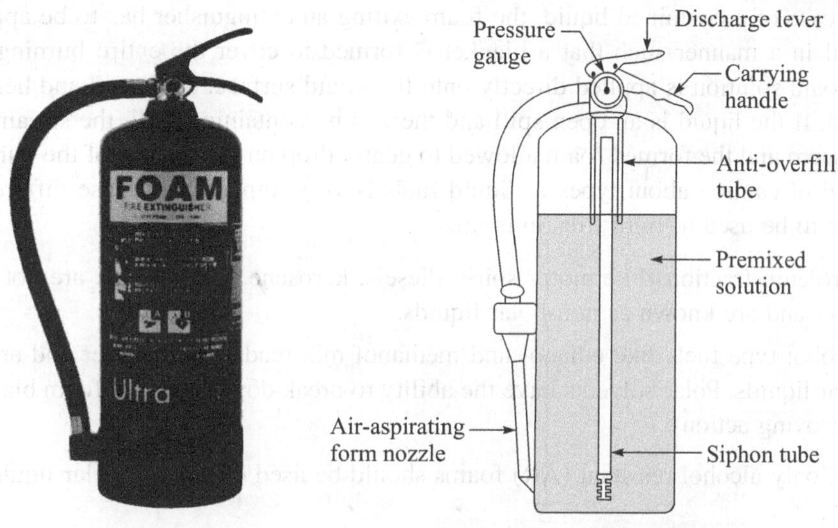

Figure 8.3 Foam extinguisher — external and sectional views.

Mechanical foam is formed by action similar to detergents (chemical foam used earlier is obsolete). Foaming agents are surfactants or surface active agents comprising chemicals that reduce the surface tension of water. Foam concentrates generally have foaming agents that are natural proteins, hydrocarbon-based surfactants (detergents) and fluorocarbon surfactants. When agitated with air, foam is formed which comprises trapped air bubbles resembling lather produced when detergent is stirred into water and can be applied to form a blanket over a burning surface.

This property is utilised to cover a burning liquid surface and thus acts by removing two faces of the fire tetrahedron — fuel and air. A special type of mechanical foam is aqueous film forming foam (AFFF), which when applied on a burning liquid surface, in addition to the foam generated, produces a thin blanket of 'water' of low specific gravity and hence can float. Thus, in addition to removal of the fuel and air faces of the fire tetrahedron, cooling effect of water is also achieved through AFFF. A manufacturer has appropriately branded its AFFF as 'light water'.

The effectiveness of a foam blanket to positively extinguish fires on burning liquids depends on factors like its

- Spreading coefficient which is indicative of ability to spread when applied on the burning liquid surface.
- Tenacity (surface tension) of the foam blanket to stay as a blanket over time.
- Drainage time or drainage rate measured as the time in minutes required for 25% of the total foam solution to drain from aspirated foam and is a measure of foam stability. Foam that has a fast drain time is normally very fluid and mobile, spreading across the fuel surface very quickly.
- Burn-back resistance which indicates the ability of a foam blanket to resist heat and flame impingement.

In the case of a contained liquid, the foam exiting an extinguisher has to be applied on the distant wall in a manner such that a blanket is formed to cover the entire burning surface. If the jet of foam solution is applied directly onto the liquid surface, the liquid and hence the fire, may spread. If the liquid is an open spill and there is no containing wall, the stream of foam is ejected upward and the formed foam allowed to gently drop on the surface of the burning liquid.

A word of caution about types of liquid fuels is very important because different types of foams have to be used to fight fires in them:

- Petroleum fractions like motor spirit, diesels, kerosene, jet fuel, etc. are not miscible in water and are known as non-polar liquids.
- Alcohol type fuels like ethanol and methanol mix readily with water and are known as polar liquids. Polar solvents have the ability to break down standard foam blankets (foam destroying action).

Hence, only alcohol resistant (AR) foams should be used on fires in polar liquid fuels.

8.4.3 Carbon Dioxide (CO_2) Extinguishers

CO_2 extinguishers are basically CO_2 gas cylinders with a non-metallic discharge horn (Figure 8.4). The discharge of CO_2 over the fire acts to displace the air in the vicinity and thus starve the fire of oxygen (smother it). However, reignition may occur if the fuel is well above its ignition temperature and the CO_2 gas has dispersed.

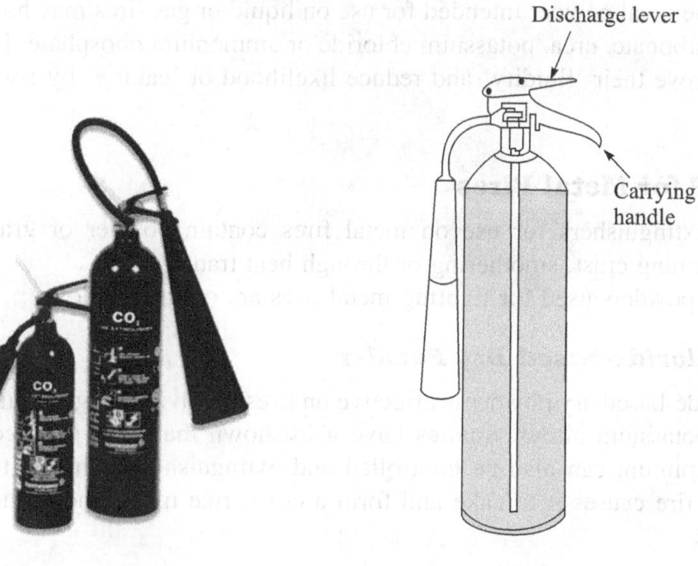

Figure 8.4 Carbon dioxide extinguisher — external and sectional views.

CO_2 extinguishers need to be periodically weighed like gas cartridges and should be replaced if the mass loss is 10% or more of the original. They are also governed by the Gas Cylinder Rules, 2004 the requirements of which have to be complied.

8.4.4 Dry Chemical Powder (DCP) Extinguishers

The shell of a DCP extinguisher has a protective coating on the inner wall. The extinguisher is fitted with a siphon tube connected to the body with a small length of high-pressure flexible hose terminating into a squeeze grip type nozzle attached to its end. The nozzle is protected with a rubber cap against entry of moisture. Dry chemical powder is filled in the body in the right quantity as per capacity (Figure 8.5).

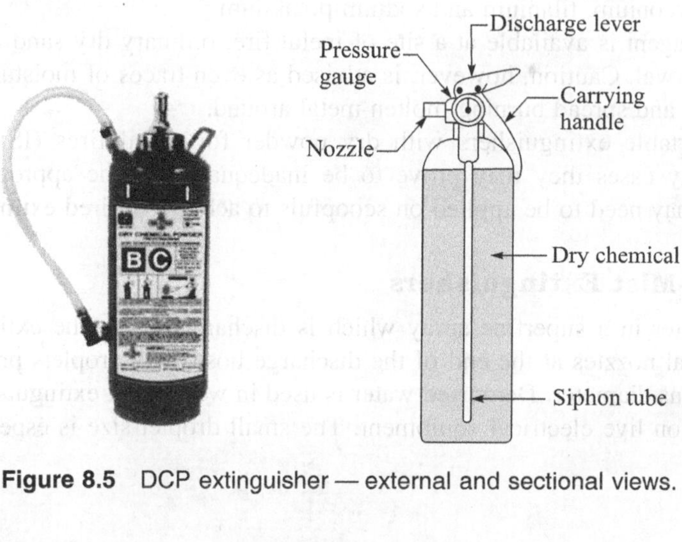

Figure 8.5 DCP extinguisher — external and sectional views.

The dry chemical powder intended for use on liquid or gas fires may have a base of sodium, potassium bicarbonate, urea, potassium chloride or ammonium phosphate. Dry powders are also treated to improve their 'fluidity' and reduce likelihood of 'caking' by adding small quantities of fine silica.

8.4.5 DCP for Metal Fires

Dry powder extinguishers for use on metal fires contain powder or granular materials that perform by forming crust, smothering or through heat transfer.

Some dry powders used for fighting metal fires are outlined below:

Sodium Chloride-based Dry Powder

Sodium chloride-based dry powder is effective on fires involving magnesium, sodium, potassium and sodium–potassium alloys. Studies have also shown that fires in zirconium, titanium and powdered aluminium can also be controlled and extinguished with this type of powder. The heat from the fire causes it to cake and form a crust, free of air and withdraws heat from the burning metal.

Copper-based Dry Powder

Copper-based dry powder smothers the fire and provides an excellent heat sink for lithium fires.

Ternary Eutectic Chloride (TEC) Powder

Based on TEC of three chlorides, viz., barium, sodium and potassium, it fuses at a low temperature to form a crust on the burning material and has high heat absorption capacity. It is suited for use in high-intensity fires (above 1200°C) in metals like sodium, potassium, calcium, aluminium, titanium, zirconium and various alloys. Barium chloride releases toxic products when applied on metal fires and the user must be protected against such products.

Graphite-based Powder

Graphite-based dry powders have been seen to be effective on fires involving high-melting point metals such as, zirconium, titanium and sodium potassium.

If no special agent is available at a site of metal fire, ordinary dry sand may be used with a long-handled shovel. Caution, however, is advised as even traces of moisture may result in a 'steam explosion' and spread burning molten metal around.

Although portable extinguishers with dry powder for metal fires (IS:11833–1986) are available, in many cases they may prove to be inadequate and the appropriate dry powder (IS:4861–1984) may need to be applied on scoopfuls to achieve desired extinguishing effect.

8.4.6 Water-Mist Extinguishers

Water mist is water in a superfine spray which is discharged from the extinguishers as mist through the special nozzles at the end of the discharge hose. The droplets produced are of the order of 25 microns diameter. Deionised water is used in water-mist extinguishers to render the mist safe for use on live electrical equipment. The small droplet size is especially effective in cooling a fire by drawing its latent heat of vaporisation. There are no ill effects on people's health too. The combined effects of small quantities of water used, almost nil water damage on assets, efficient cooling achieved, no adverse health effects and safe application on food as well as live electrical systems has made water mist a very preferred fire fighting medium.

8.4.7 Clean Agent Extinguishers

In the 1980s and up to the mid-1990s, Halons were a most preferred fire extinguishing medium due to its numerous positive aspects. However, concerns about their ozone depleting potential (ODP) and stipulations of the Montreal Protocol brought an end to their use though Halon 1211 is still permitted to be used on aircraft. Halon alternatives are being researched and some products such as Halotron and inert gas-based products are available in the market.

8.4.8 Modular Automatic Extinguishers

A useful innovation of extinguishers is the modular automatic extinguishers which contain a design quantity of extinguishing medium in a sealed container. The seal has a fixed temperature detector and the module may be mounted on the ceiling or a wall (Figure 8.6). When the detector senses the design temperature, the seal opens and the medium is discharged in the area. Such modules have been installed in areas of housing sensitive, high-value or critical assets.

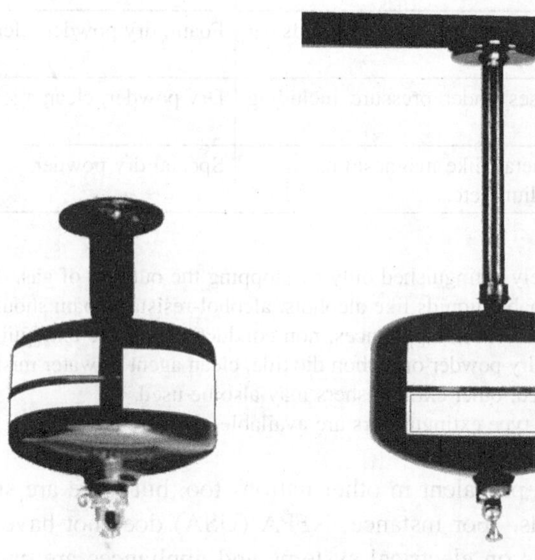

Figure 8.6 Typical modular extinguishers.

- Automatic extinguishers may contain DCP or clean agents in modular containers fitted with fixed temperature detectors and discharge nozzles. The upper end has fixtures for mounting on the ceiling or wall.
- A pressure gauge is fitted to monitor availability of adequate working pressure.

Advantages of modular automatic extinguishers

- Detects and extinguishes fire quickly and efficiently.
- Provides concentrated fire protection for vital equipment.
- Cost-effective compared to automatic fixed fire protection systems.
- Easy to install and relocate.
- Not dependent on electrical power supply.

8.5 CLASSIFICATION OF FIRES AND SUITABILITY OF EXTINGUISHERS

Primarily as an aid to select the right type of extinguisher, IS:2190–2010 has classified fires on the basis of material that is burning as is explained in Table 8.2.

Table 8.2 Classification of fire for extinguisher selection

Class of fire	*Materials involved*	*Type of extinguisher*
A	Solid organic combustibles like wood, paper, plastic, etc.	Water, foam, ABC dry powder and halocarbons
B	Flammable liquids, liquefiable solids or similar fuels	Foam, dry powder, clean agent and CO_2
C	Flammable gases under pressure including liquefied gases	Dry powder, clean agent and CO_2
D	Combustible metals like magnesium, aluminium, sodium, etc.	Special dry powder

Notes

- Gas fires can be positively extinguished only by stopping the outflow of gas.
- For fires on polar flammable liquids like alcohols, alcohol-resistant foam should be used.
- For fires involving live electrical appliances, non-conductivity of the extinguishing media is vital and extinguishers expelling dry powder or carbon dioxide, clean agent or water mist may be used. Once the equipment is de-energised, other extinguishers may also be used.
- Even though water-mist type extinguishers are available, they are not referred in IS:2190-2010.

Similar classification is prevalent in other nations too, but there are significant differences among the various standards. For instance, NFPA (USA) does not have a class for gas fires and 'electrical fires' or fires on electrical systems and appliances are categorised as Class C. Another class of fire is K (kitchen) in USA and F (fat) in the UK, both of which refer to fires in the kitchen involving fats and edible oils.

The selection of extinguishers based on fuel type as given in Table 8.1 is based on the principles by which the different types of extinguishers function to achieve extinguishment of fires. These principles are listed in Table 8.3.

Table 8.3 Principles of extinguishment

Medium	*Working principle*
Water	Apply at base of flame to cool fuel to inhibit pyrolysis and vaporisation. Must not be used on live electrical equipment.
Foam	Blankets (smothers) liquid surface, inhibits fuel vaporisation, cuts off contact with air.
Carbon dioxide	Displaces air to remove contact of fuel vapours with air; can reignite if fuel is heated to above ignition temperature.
Dry powder	Inhibits flame chain reaction; can reignite after the powder is dissipated if the fuel is heated to above ignition temperature.
Halon	Inhibits flame chain reaction.
Dry powder (metal)	The heat of the fire causes the dry powder to melt and form a crust that excludes air and withdraws heat from the burning metal. Specific powders are suited to fires in different metals and alloys.
Water mist	Draws latent heat of vaporisation from the fire and the discrete globules of deionised water offers safety during application on live electrical systems.

Table 2 of IS:2190–2010 and Table 8.4 lists the respective Indian standards (specifications) for the respective extinguishers, the fire extinguishing media contained and their suitability for use on different classes of fire.

Table 8.4 Specifications and suitability of different extinguishers

Type of extinguisher	*IS*	*Class of fire*			
		A	B	C	D
Water type (gas cartridge)	13385	✓	×	×	×
Mechanical foam type (gas cartridge)	13386, 4989 (or ISO 7203)	✓	✓	×	×
Mechanical foam type (stored pressure)	14951, 4989 (or ISO 7203)	✓	✓	×	×
Dry powder type (gas cartridge)	10658, 4308	✓	✓	✓	×
Dry powder type (stored pressure)	4308	✓	✓	✓	×
Dry powder for metal fires	11833, 4861	×	×	×	✓
Carbon dioxide type	2878, 8149	×	✓	✓	×
Halon 1211 type	4862 (Part 1) & ISO 11108	✓	✓	✓	×

Note: Fire buckets have traditionally been considered as first-aid fire fighting appliance. However, as per IS:2190–2010, they are not intended to be alternates for water-type extinguishers and may be installed as supplementary provisions. The TAC Fire Protection Manual, on the other hand, equates fire buckets with water-expelling extinguishers.

8.6 RATING EXTINGUISHERS

The formulation of IS:15683–2006, Specification for performance and construction of portable fire extinguishers added the new dimension of rating extinguishers to indicate the size of a fire a particular extinguisher can combat in line with practice in developed nations. It is based on the amount of extinguishing medium used to extinguish a fire of given size under conditions of test. This calls for labelling extinguishers both with the class of fire as well as the size of fire and is applicable to fires of Class A and Class B. Metal fires (Class D) are not classified by size but the medium used must be appropriate for the metal involved.

Class A test fires are set up in the form of cribs consisting of specific timber (*Cryptomeria Japonica*, a type of deodar) pieces of prescribed size and tiers. Fire sizes are rated from 1A to 6A with increasing number of member pieces and tiers in the crib. Similarly, Class B fires are rated from 8B to 34B depending on the pan diameter and hence the surface area of the burning fuel. 8B to 13B fire tests are conducted indoors and larger fire tests are conducted outdoors but with wind speeds under 3 m/s. The fuel generally used is n-heptane.

The minimum effective discharge time of extinguishers rated 1A should be 8 seconds, while it should be 13 seconds for extinguishers rated 2A or higher. For all categories of B extinguishers, the minimum discharge time is 8 seconds but their throws are different.

8.7 HAZARD CATEGORISATION AND PLACEMENT OF EXTINGUISHERS

Extinguishers are required to combat fires in their early stages whether on the building components or the contents. IS:2190–2010 provides guidelines for classifying various occupancies as light (low), ordinary (medium) and extra (high) in terms of their potential fire severity due to contents and are intended to help in determining the type and number of extinguishers for a particular location. Table 8.4 provides a summary of extinguisher requirement for various hazard categories based on the guidelines provided in annexure to the standard. Importantly, IS:2190–2010 also recommends the installation of rated extinguishers in Class A and Class B areas wherein density of extinguishers and recommended travel distances are in variance with the recommendations included in Appendix 3.1.

As per the TAC fire protection manual, for any property, the basic protection should be appliances suitable for Class A fires. Thus, initially, water-type extinguishers are proposed on the recommended scale. Thereafter, if fires of other classes are anticipated, an appropriate number of appliances suitable for anticipated fire types are selected to replace the ones proposed for Class A fires. The manual also stipulates the provision of fire buckets (water and sand) and also permits them as replacements for extinguishers.

For protection of electrical equipment, TAC stipulates the following:

- For compartments housing electrical equipment only, at least two 2 kg dry powder or carbon dioxide extinguishers should be provided within 15 m of the apparatus.
- Where motors and/or other electrical equipment are installed in rooms that have other equipment also, one 5 kg dry powder or carbon dioxide extinguisher may be installed in addition to the requirement of Class A and Class B fire protection. The same extinguisher may be deemed to afford protection to all apparatus within 15 m thereof.

It should however be remembered that the recommendations in codes are the minimum and the actual types, number and location of extinguishers should be based on a study of the premises, activities and location of specific hazards.

8.8 CONCLUSION

- To use an extinguisher the fire must be discovered while it is still small.
- The person using an extinguisher must be trained to use it properly. Speed is important.
- Extinguishers should be located for visibility and easy access preferably near the entrance.
- Post signs to show the locations of fire extinguishers, particularly in large floor areas. The signs should be large enough to be seen clearly from a distance.
- Extinguishers must not be blocked by equipment, hanging clothes or other objects.
- They should be installed at places where work movements such as of trucks and cranes will not damage them.
- If located outdoors, they should be protected against the weather.
- Select and use the right type of extinguisher.
- Get within effective range but stay safe.
- When using an extinguisher, ensure that there is an escape way just in case you have to retreat.
- In the case of hazardous material stores, locate extinguishers outside. This will force the user to exit the room and then decide whether to re-enter to fight the fire.
- Extinguishers should be maintained and inspected regularly for completeness of parts, corrosion, loose fittings, clogged discharge, visible damage, available pressure, etc.
- On vehicles or in areas where extinguishers are subject to jarring or vibration mount extinguishers on brackets designed to withstand vibration.

- The powder in DCP extinguishers should be checked for caking and sieved /refilled as required.
- Fire extinguishers should be hydro tested periodically as recommended by manufacturer/ codes of practice.
- It is advisable to recharge all extinguishers immediately after use regardless of how much was used.

QUESTIONS FOR REVISION

1. Why are portable extinguishers called "first aid" appliances?
2. What are the ways in which the contents of extinguishers are expelled?
3. What are the different ways in which water as an extinguishing medium is delivered from extinguishers? Contrast them in terms of throw length and manner of discharge.
4. Water is the primary content of water, foam and water-mist extinguishers. What are the differences in the discharge nozzles of the three types?
5. How are fine particles of water more effective in cooling than full bore streams?
6. What are the differences in dry powder used for fighting Class C and Class D fires? Can they be used as substitutes for each other? Give reasons for your answer.
7. Can fire buckets be replacements of water-type extinguishers? Support your answer with reference to IS 2190 and the TAC Fire Protection Manual.

9

Fixed Fire Protection Systems

Failure to prepare means preparing for failure.

—Anon

9.1 INTRODUCTION

It is best if all fire outbreaks are prevented. Being able to extinguish them at an early stage by using extinguishers comes next. Yet there are situations where fires start big or have grown big where larger appliances like fire tenders or fixed protection systems like those for fighting fires in storage tanks have to come into action. This chapter describes the working principles of some fixed fire protection systems based on water and other extinguishing media.

9.2 WATER-BASED FIXED FIRE PROTECTION SYSTEMS

Water in various forms has been (and still is) the most preferred fire fighting medium because of its availability, ability to draw heat due to its high heat capacity and the ease with which it can be conveyed over long distances. When broken up into fine droplets, its high latent heat of vaporisation makes it a wonderful heat-withdrawing medium. Water is also used to protect structures against damage by exposure to high incident heat flux. Another use of water in fire protection is when it is mixed with some liquid hydrocarbon fuels to render them non-flammable either through emulsification or by dilution. Water is also an essential component of foam systems which is a basic protection for hydrocarbon fires. In this section, we will discuss some water-based fixed fire protection systems.

9.2.1 Small Bore Hose Reel

Small bore hose reels, as the name implies, are rubber or plastic hoses of small bore (generally 19–32 mm internal diameter) reeled up on drums mounted on walls in different parts of a building where the reel may either be fixed or can be rotated through 180° on the vertical plane (Figure 9.1). The hose is connected to the water supply of the premises and the discharge end has a nozzle with provision for shutting off. In case the available water pressure is not adequate, the hose reel may source its water from a hydrant riser with appropriate pressure reducers. Throw from the nozzle should be at least 6 m, which means that the nozzle at the delivery end should reach within 6 m from any point in the building.

Figure 9.1 Hose reel.

Small bore hose reels are meant for use by occupants of a building as first aid device to fight fires inside a building. With the exception of small dwellings and low rise educational buildings, NBC 2005 recommends the provision of small bore hose reels in all other occupancy groups. IS:884–1985 is the specification for small bore hose reels and IS:3884–1989 is the relevant code of practice. The TAC Fire Protection Manual specifies some performance requirements. There are minor variations between the requirements of IS:884–1985 and the TAC Fire Protection Manual (e.g., minimum water flow rate of 24 and 22.5 lpm respectively).

Hose reels are mounted as standard equipment on some types of fire tenders and are also widely used on board ships.

9.2.2 Hydrant Systems

A fire hydrant system is essentially network of pipes connected to a source of water supply and provided with outlets for tapping water under pressure with provision to deliver water in desired discharge patterns of stream, spray or fog through attached hose and nozzle with intent to effectively extinguish the fire and/or protect endangered assets or people (both occupants and fire fighters). Also integral to the system are pumps to achieve desired flow and pressure, isolation valves on pipelines to isolate sections when required for operation or maintenance purpose and outlets distributed over the protected premises with outlet valves and hose connectors (Figure 9.2). The hydrant mains (as the piping is called) or network is effective in reaching water to all areas of the premises and is also used to serve other fire protection needs such as monitors, foam delivery and application systems, exposure protection spray systems, indoor sprinkler systems, etc.

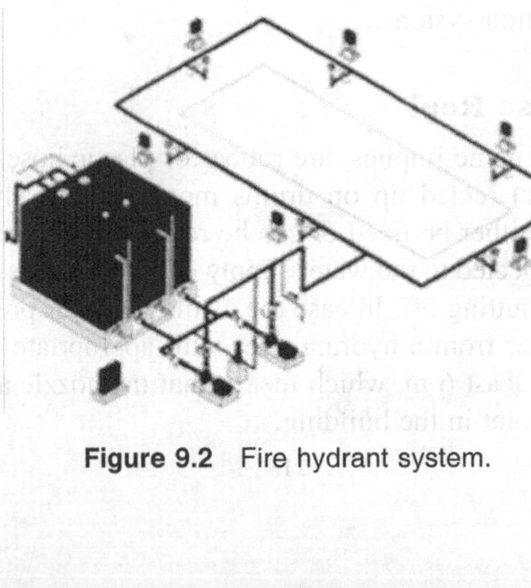

Figure 9.2 Fire hydrant system.

The system may be designed for and installed within or outside buildings and structures distributed across the protected premises. Large industries design and install their own private fire hydrant networks spread all over their premises. Other occupancies such as hotels, hospitals, commercial buildings, shopping malls and airport terminals too have their hydrant and riser systems. Industrial townships of large organisations have their own hydrant systems. However, only some Indian cities have public fire hydrant systems unlike most urban areas in the developed countries. A well-designed and laid out fire hydrant system is of primary importance in fighting major fires. It can be used to

- cool burning materials and extinguish the fire.
- keep nearby combustible materials cold to prevent ignition and spread of fire.
- protect adjacent assets from damage through thermal radiation by cooling them.
- provide protection against thermal radiation for fire fighters.

Irrespective of its location, the components of a hydrant system are essentially the same. These include:

- A source of water supply
- Pumps
- Hydrants mains (risers for elevated locations) and valves
- Hydrants
- Hoses and nozzles
- Wheeled or ground monitors (depending on requirements)

A public or municipal hydrant system is laid out to cover various sections of an urban area. It is a much more complex system than a private system installed to protect either one or more buildings in a group or an industrial or business facility. The rest of our discussion on hydrant systems will, therefore, be confined to private systems, which may be considered as the heart of active fire protection in many industrial and commercial properties.

NFPA 14, Standard for Installation of Standpipe, Private Hydrant and Hose Systems is a widely followed code for hydrant systems. The significant Indian standards on hydrants are:

- IS:3884–1989, Code of practice for installation and maintenance of internal fire hydrants and hose reels.
- IS:13039–1991, Code of practice for external hydrant systems — provision and maintenance.
- IS:9668–1991, Code of practice for provision and maintenance of water supplies and fire fighting.
- IS:908–1975, Specification for fire hydrant, stand post type.
- SP 7–2005, National Building Code of India 2005.
- Model Building By-laws, Ministry of Urban Development, Government of India and local by-laws.
- Fire protection Manual, Tariff Advisory Committee.
- OISD 116, Fire Protection Facilities for Petroleum Refineries and Oil/Gas Processing Plants.
- OISD 117, Fire Protection Facilities for Petroleum Depots, Terminals, Pipeline Installations and Lube oil installations.

Sources of Water Supply

Water to a fire hydrant system may be either from a natural reservoir, a perennial stream or the sea or a man-made static tank or over ground reservoir. If the source is a stream or the sea, it may be advisable to store the water in a dedicated reservoir on site. If sea water is used, the reservoir, pumps, piping materials, fittings and applicators (such as sprayers, monitors, etc.) should be suited for such use. In case of reservoirs, its required capacity will be decided on the basis recommended in standards. For buildings, different Indian standards recommend different locations (underground or on the terrace), capacities and replenishment rates. Hence, while designing a system, requirements of the reference code should be followed. Where building by-laws exist, stipulations therein must first be complied; additional provisions may be made based on engineering judgement and other practical considerations.

In high hazard industries like the hydrocarbon sector, standards like those of Oil Industry Safety Directorate (OISD) recommended that there should be adequate water supply to fight two major simultaneous fires for 4 hours. In the case of large fire-water storage requirements, standards recommend that the total capacity be divided into two interconnected compartments of equal capacity to facilitate cleaning of the reservoirs without disabling the hydrant system.

Fire Water Pumps

Water from the source needs to be transported through the pipe network for delivery at the hydrant outlets and other end-use points. This function is performed by pumps and may rightly be called *the heart of a hydrant system*. Standards stipulate the discharge flow and pressure at outlets and pumps are selected to deliver these parameters.

Pumps for hydrant systems may be either electric or diesel powered. Electrical pumps require a dedicated feeder such that in case electrical power to the premises has to be cut off under fire conditions, electric pumps can continue to work. However, as a further precaution, standards stipulate that at least half the pumps are diesel powered. It is also required that the fire-water pumps are installed in a pump house remote from the areas that may be affected by fire (Figure 9.3).

Figure 9.3 Fire-water pump house.

Hydrant Mains and Risers

The pipe networks from the pumps to the hydrants or monitors are known as *mains.* It is advantageous to lay out mains in grids rather than in branched pattern for better flow and pressure drop considerations. Grid layout also permits isolation of sections (if required) without disabling network availability at any time. Hydrant mains are also provided with shut-off valves so that sections may be isolated for operational reasons or for maintenance. For convenient inspection and maintenance, hydrant mains are installed over ground with adequate supports. However, they should be buried and appropriately protected against damage at road or railway crossings, at places where heavy equipment or vehicles may be required to enter or where there is potential for severe ground fires. When the hydrant system is required to serve at higher elevations, tappings from the mains are taken up through pipes known as *risers*. A rising main is a pipe installed vertically in a building or elevated work structure connected to the fire water mains at grade and outlets with shutoff valves (***landing valves***) at various levels throughout the building or structure. There are two types of rising main:

- Wet riser and
- Dry riser

In a ***wet riser,*** the rising main is kept charged with water under pressure, which is immediately available for use on any upper floor at which a hydrant (or landing valve) is installed. The same duty is served by what is known as a ***down-comer*** in which a pipe is connected to a roof-top tank.

In a ***dry riser***, the rising main is normally kept empty and water to it is supplied from water tenders through inlets located on the building structure near grade or in covered boxes underground. Thus it is meant for use by firemen.

Hydrant Outlets, Monitors and Landing Valves

Water from the hydrant system is drawn for end use at

- hydrant (stand post or underground) outlets (Figure 9.4)
- monitors (fixed and mobile) (Figure 9.5) anding valves at elevated levels (Figure 9.6) or
- fixed foam systems
- cooling water systems

The outflow from a hydrant outlet could be used through hoses and nozzles either on a fire to attempt extinguishment or for cooling nearby assets (both structures and contents), to make foam solution with foam compound using a foam making branch or to feed mobile monitors through hoses. Outlets at stand posts and landing valves have quick fitting female coupling with a lug to lock the male coupling of the hose.

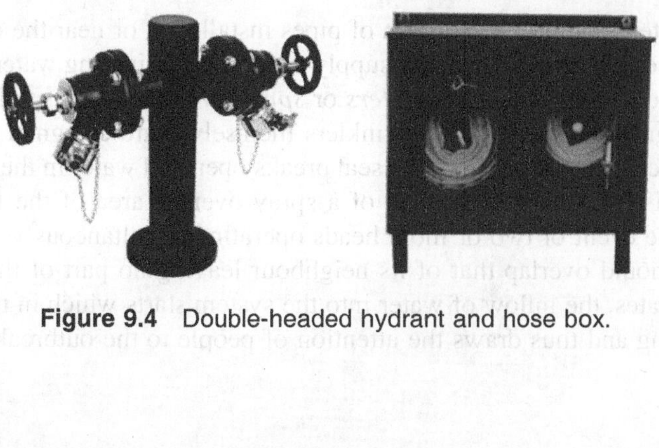

Figure 9.4 Double-headed hydrant and hose box.

Figure 9.5 Water monitor.

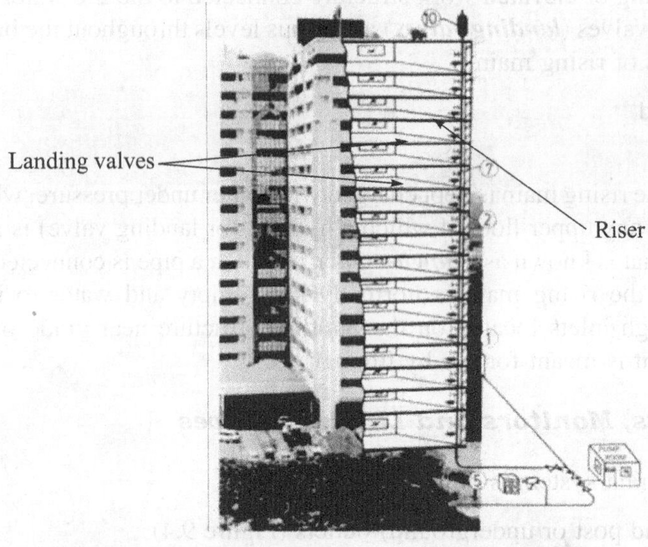

Figure 9.6 Riser system with landing values.

9.3 AUTOMATIC SPRINKLERS

Automatic sprinklers are devices which can detect a fire automatically and then act to control (and possibly extinguish it) with minimal losses. Automatic sprinkler systems have been used for well over a century and have done yeoman service in fire scenarios in buildings throughout the world. Their absence has cost lives as in the case of MGM Grand Hotel in Las Vegas in November 1980.

A sprinkler system comprises a system of pipes installed at or near the ceiling of a building compartment connected to a reliable water supply capable of delivering water at desired flow rate and pressure. Sealed devices called *sprinklers* or *sprinkler heads* are installed on the pipes and spaced as per design (Figure 9.7). The sprinklers themselves are designed such that, when the ambient temperature rises to a set limit, the seal breaks open and water in the pipes is discharged through the opened sprinklers in the form of a spray over an area of the floor. Sprinklers are so spaced that in the event of two or more heads operating simultaneously, the discharge circle of each sprinkler should overlap that of its neighbour leaving no part of the floor unprotected. Once the head operates, the inflow of water into the system starts which in turn rotates a turbine causing a bell to ring and thus draws the attention of people to the outbreak of fire.

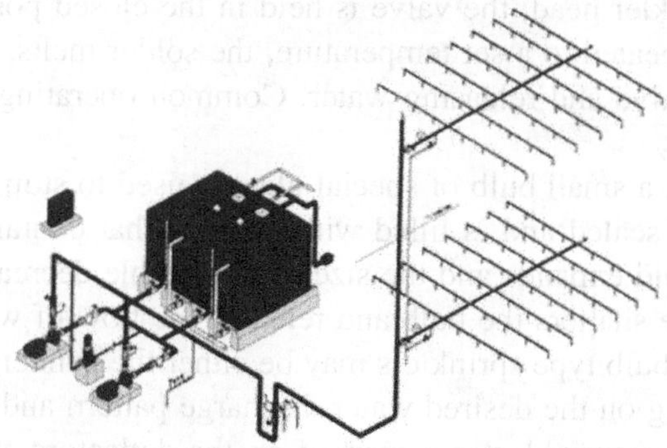

Figure 9.7 Sprinkler system.

9.3.1 Sprinkler Heads

Sprinkler heads are of different designs but they may be divided into two categories:

- Those which operate by the melting of a solder joint (Figure 9.8) and
- Those in which a bulb is ruptured by expansion of a contained liquid and air bubble (Figure 9.9).

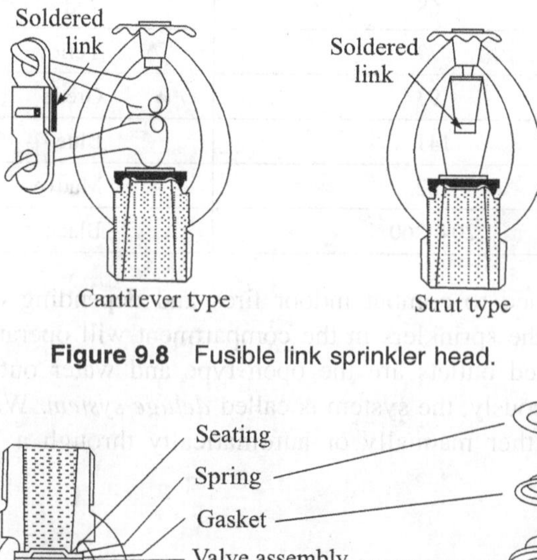

Figure 9.8 Fusible link sprinkler head.

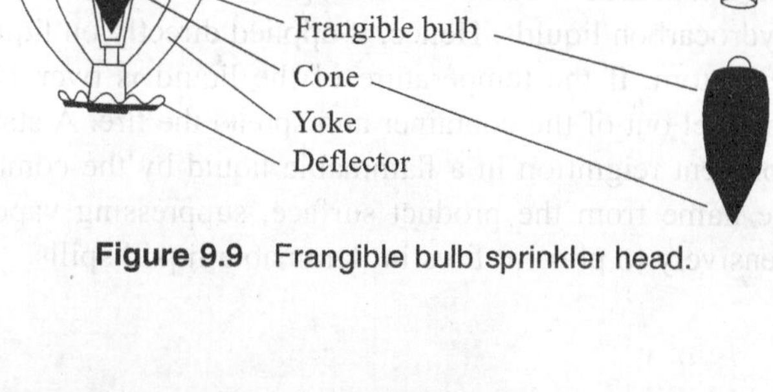

Figure 9.9 Frangible bulb sprinkler head.

In a solder type sprinkler head, the valve is held in the closed position by two metal parts soldered together. When heated to a set temperature, the solder melts, allowing the parts to fall apart, thus opening the valve and releasing water. Common operating temperatures are 68, 93 and 141°C.

In the bulb type head, a small bulb of special glass is used to stop the water flow. The bulb is hermetically (air tight) sealed and is filled with a liquid that contains an air bubble. As the temperature rises, the liquid expands and the size of the bubble decreases until it disappears. A further rise of temperature shatters the bulb and releases the flow of water.

Both solder type and bulb type sprinklers may be either the pendent type or the upright type and are selected depending on the desired water discharge pattern and protection.

Temperature ratings of sprinklers are marked on the deflectors and the liquid in the bulb is coloured specific to its operating temperature (Table 9.1). IS:15105-2002 stipulates seven temperature levels with respective colours of contained liquid. The liquid employed is non-freezing and the bulb itself is not subject to corrosion.

Table 9.1 Sprinkler Liquid Colours (Source IS:15105–2002)

Sprinkler operating temperature (°C)	*Colour of liquid in bulb*
57	Orange
68	Red
79	Yellow
93	Green
141	Blue
182	Mauve
204–260	Black

A sprinkler is designed to combat indoor fires and depending on the characteristics of the fire, one or more of the sprinklers in the compartment will operate. However, if designed in a way that the installed outlets are the open type and water outflow through all outlets can be actuated simultaneously, the system is called *deluge system.* Water flow into the system can then be activated either manually or automatically through a linked detection system (Figure 9.10).

9.4 FOAM SYSTEMS

Foam has for long been a favourite medium for extinguishing fires in flammable liquids. Water is heavier than most hydrocarbon liquids. Hence, if applied directly on liquid hydrocarbon fires, water will sink to the bottom. If the temperature of the liquid is over 100ºC, the water may even boil and throw the fuel out of the container and spread the fire. A stable foam blanket can extinguish a fire and prevent reignition in a flammable liquid by the combined mechanisms of cooling, separating the flame from the product surface, suppressing vapours and smothering. Foam is also used extensively to prevent fires in flammable liquid spills.

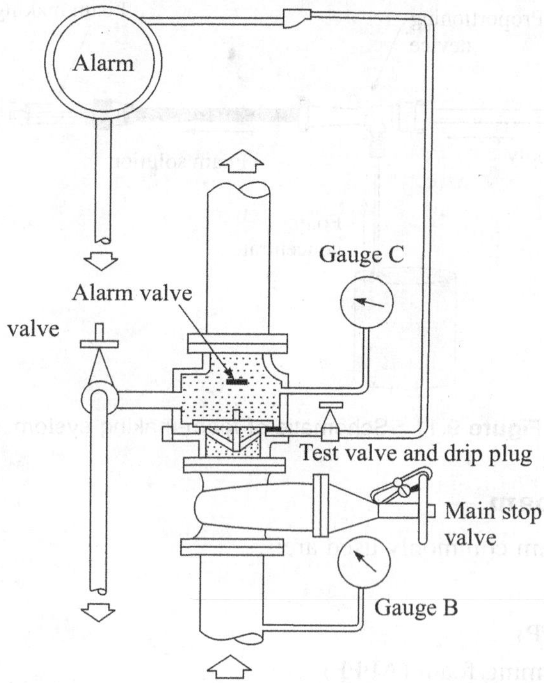

Figure 9.10 Sectional view of wet-pipe sprinkler valve.

Before reviewing the merits of the different types of foam concentrates, there are certain terminologies associated with foam that must be understood.

Fire fighting *foam* is a stable mass of small air-filled bubbles produced from three ingredients—water, foam concentrate and air to form a homogeneous foam blanket. It is very similar to the lather produced when a detergent is mixed in water and then agitated.

$$\text{Foam concentrate} + \text{Water} \longrightarrow \text{Foam solution}$$

$$+ \text{Air} \longrightarrow \text{Finished foam}$$

Dilution, ***mixing*** or ***proportioning rate*** of a foam concentrate is the ratio of foam concentrate and water mixed to make foam solution (expressed as percentage of concentrate in solution). Concentrations widely used are 3% and 6% and IS:4989–2006 refers to these as well as 1% concentrations. Different grades of concentrate must be marked appropriately and not mixed. Foam-making equipment is also designed based on concentrate quality (Figure 9.11).

Expansion ratio is the volume expansion of foam solution to finished foam. Thus, an expansion ratio of 10 would mean that one litre of foam solution would, after aeration, expand to become 10 litres of finished foam. Based on the expansion ratio, foams are rated as ***low*** (1: 2 to 20), ***medium*** (1: 20 to 200) and ***high*** (1: 200 to 1000 or more) expansion as seen from different sources of literature.

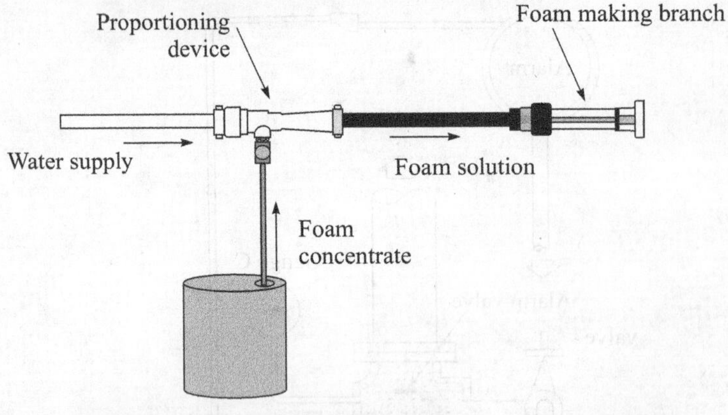

Figure 9.11 Schematic of foam-making system.

9.4.1 Types of Foam

Types of mechanical foam commonly used are:

- Protein
- Fluoro-protein (FP)
- Aqueous film-forming foam (AFFF)
- Film forming fluoro-protein (FFFP)

Other mechanical foams also used for specific uses:

- Alcohol-resistant AFFF (AR-AFFF)
- Synthetic — medium or high-expansion types (detergent based)
- Class "A" foam concentrate
- Wetting agent

Protein foam concentrate based on hydrolysed protein, foam stabilisers and preservatives (available in 3% and 6% concentration) produces very stable foam. Protein foam must always be used with an aspirating type discharge device. The application technique for protein foam is critical: the foam should be applied as gently as possible to the flammable liquid surface or else it may get contaminated and lose its effectiveness.

Fluoro-protein foam (available in 3% and 6% concentration) is protein foam with added fluorocarbon surfactants. The addition of these surfactants in the concentrate makes the fluoro-protein foam more resistant to fuel contamination and the foam blanket more mobile when discharged onto the flammable liquid.

Aqueous film forming foam (AFFF) concentrate (available in 1%, 3% and 6% concentration) is made from synthetic materials such as:

- Hydrocarbon or fluoro chemical surfactants
- Small amounts of salts
- Foam stabilisers

A plus point of AFFF over protein and fluoro-protein foams when used on liquid hydrocarbon fires is that it forms an aqueous film on the surface of the burning liquid as the foam solution drains from the foam blanket. Another advantage is that it is suitable for use with dry chemical extinguishing agents.

Film-forming fluoro-protein (FFFP) is a derivative of AFFF and fluoro-protein. FFFP concentrates have the quick knockdown of AFFF and the added burn-back resistance of normal fluoro-protein foam. However, FFFP does not have the quick knockdown of the AFFF's when used on a spill fire such as an aircraft crash or a highway spill and in deep fires they do not have the burn back resistance of fluoro-protein.

Synthetic/detergent (high expansion) foam concentrate (normally between 1.5% and 2.5% concentration) is made from hydrocarbon surfactants and solvents and used through devices such as high expansion type foam generators. It is useful in places like basements, mine shafts, ship's holds, tunnels or aircraft hangars where volume fire control is required.

Class "A" foam concentrate is a mixture of foaming and wetting agents. The former allows water to remain and cling to vertical and horizontal surfaces without run off and the latter permits greater penetration into fires of class "A" fuels and greater withdrawal of heat. A small quantity of class "A" foam concentrate added into a water stream can increase the effectiveness of the water several fold.

The term *shelf life* of a foam concentrate describes the length of time for which unused concentrate remains stable and usable. It depends on factors like the composition of the concentrate, the ambient temperature, container materials and if it is stored in its original container. A shelf life of 10–15 years is possible for AFFF if the recommended conditions of storage are maintained. The common shelf life of protein concentrates is 2 years.

9.4.2 Foam Usage

Foam is an essential item for fire protection in hydrocarbon storage, handling and processing installations and guidelines like the OISD standards recommend foam compound inventories to be maintained on site (Figure 9.12). The standards also contain guidelines for foam application rates and foam pourers in cone roof tanks, floating roof tanks (rim seal protection) and floating roof tanks with sunken roof, dyke application, etc.

High-expansion foams are useful for rapid flooding of large volume spaces like aircraft hangars, special types of warehouses and use on airport runways. In the last instance, crash tenders may be used to deliver the foam (Figure 9.13).

9.5 WATER SPRAY SYSTEMS

The term *water spray* refers to the discharge of water in a predetermined pattern, droplet size, velocity and density from specifically designed nozzles or other devices. They are usually applied in special fire protection requirements since the protection can be designed to provide for effective fire control, fire prevention or exposure protection. However, not all water spray systems can be used for all types of protection desired and end-use must be ascertained before a system is chosen. These systems may be independent of or supplementary to other forms of protection.

(a) (b)

Figure 9.12 Foam maker and foam pourer: (a) for fixed roof tank (b) for floating roof tank rim seal protection.

Figure 9.13 High expansion foam system for hangar protection.

A water spray system comprises fixed pipes connected to a reliable water source and equipped with water spray nozzles for designed water discharge distributed over the area to be protected. An automatically or manually actuated quick-opening ***deluge valve*** is located between the water source and the piping system to start the flow of water. Automatic actuation is achieved through detectors installed parallel to the water spray nozzles on a pneumatic system.

There are two types of systems viz., high velocity (HV) and medium velocity (MV) water spray systems. The spray nozzles are open water discharging devices designed to discharge water under pressure over an area in a predetermined manner. By convention, nozzles used for HV systems are called *projectors* and those used for MV systems are called *sprayers*. Both types of nozzles are available in a range of orifice sizes and discharge angles. The piping network for a spray system will have vertical mains and horizontal rings and discharge nozzles should be installed only on the latter.

9.5.1 High-Velocity Water Spray System

The use of water to fight oil fires is considered a taboo. However, the use of HV water spray on oil fires converts the flammable liquid temporarily into an emulsion which cannot burn.

A special type of nozzle, called a *projector,* is used to discharge a cone of water in the form of evenly distributed broken streams of high velocity and momentum. HV spray nozzles produce water droplets of diameter 1.5–2.5 mm with high velocity at an operating pressure range of 3.0–6.0 kg/cm^2(g). The impact of the discrete high-velocity water particles breaks up the oil into tiny globules which form an emulsion with the water and the emulsion cannot burn. Emulsification of the oil also induces cooling, which reduces vaporisation.

Oil–water emulsions that may be formed ordinarily separate out into oil and water in a relatively short time. Emulsions created by HV systems on the other hand are sustained long enough to prevent reignition. HV systems can be used in both mineral and vegetable oil as also in animal fat fires. They are installed to extinguish fires in liquids with flash points of 65°C or higher and the three principles of extinguishment the system employs are simultaneous emulsification, cooling and smothering. The result of applying these principles is to rapidly extinguish the fire.

Liquids that form stable emulsion include:

- **Mineral oils** such as lubricating oil, transformer oil, switch oil, light diesel oil, gas oil and fuel oil.
- **Vegetable oils** such as turpentine, cotton seed oil, soyabean oil, linseed oil, castor oil, olive oil and coconut oil.
- **Animal oils** such as whale oil, cod-liver oil and halibut oil.

HV water spray system is used in paint and varnish processes, vegetable oil extraction plants and waterproofing factories but the most widely used location is electricity generating and distribution stations, for the protection of oil-filled transformers and switchgear and the lubricating system of turbo alternators (Figure 9.14). They have also been used in power plants on lubricating oil systems, seal oil systems, turbo-alternators, etc. as localised or 'spot' protection for the specific equipment.

(a)

(b)

Figure 9.14 (a) HV diffuser nozzle and (b) transformer protection.

HV water spray systems have been used to extinguish fires in light mineral oils such as petrol, kerosene and benzene but unlike with heavier oils, emulsions with the light oils are transient.

Hence, complete extinguishment of fire in light mineral oils cannot be assured unless the whole of the burning surface is bombarded simultaneously from projectors.

HV spray systems are widely used to control/extinguish fires on liquid fuels of flash points higher than 65°C. They are also recommended for exposure protection where nozzles-to-target distances are long particularly in windy external areas and where high-density water application is required.

9.5.2 Medium-Velocity Water Spray System

MV water spray systems are widely used to protect vessels, equipment and structures against exposure to heat from fires in the vicinity or adjoining areas that could result in damage to or distortion of target objects. Cooling also helps to control pressure rise in vessels that could otherwise rupture and lead to release of contents and fire spread. Design of the system should take care that spray coverage is complete with some overlap of water discharged from adjacent sprayers such that 'hot spots' are not formed. In addition to the vessel surfaces, protection should also be provided to supports on which the vessel is mounted (Figure 9.15).

Figure 9.15 MV water spray nozzle and tank spray protection.

MV water spray systems can be used to extinguish fires in polar liquids by dilution. They are also used for protection against fires in coal conveyors, cable galleries, etc. MV water spray systems have also been used to control burning in spill fires involving flammable liquids with flash points below 65°C and effectively reducing the rate of burning.

It is reported that exposure protection may be used to contain incident thermal radiation within the maximum allowable heat input of 16,290 kcal/h/m^2. Similarly, the tank shell temperature which shall not preferably exceed 343°C, can be brought down to 100°C by water spray system.

MV spray nozzles produce water droplets of diameter 0.4 mm or less discharged at a velocity of 15–20 m/s. Larger the discharge angle, smaller is the droplet size. MV sprayers are primarily used for exposure protection.

9.5.3 Water Application Rates

Recommended water application density in the case of spray systems depends on factors such as the type of protection and its working principles, the protected target etc. and ranges from 1 to

12.2 lpm/m^2 at pressures ranging from 140 to 600 kPa. For details on recommended discharge densities, pressures and flow velocities in feeder lines, one may refer to relevant codes of practice.

9.5.4 Limitations

Water spray systems are used extensively in industry but, like all other provisions, have their own limitations such as

- When used on fires in non-polar liquid fuels that are at temperature higher than the boiling point of water, frothy oil–water may be released and even lead to spread of fire.
- When cold water is applied on very hot metal surfaces, they may be subjected to thermal shock and resultant damage or distortion.

9.6 WATER MIST SYSTEMS

It has been reported that precursors to today's water mist systems appeared in the 1950s but it was in the 1990s that they got to be better known particularly because those were the years when halogenated hydrocarbon systems were being phased out and the hunt for alternative clean agents was hot. The term *water mist* has been defined as fine water sprays in which 99% of the volume of the spray is in drops of diameter 1000 microns or less (Figure 9.16). In contrast to other water-based fire protection systems such as sprinklers, MV water spray and HV water spray systems, in the case of water mist, the very fine water mist globules:

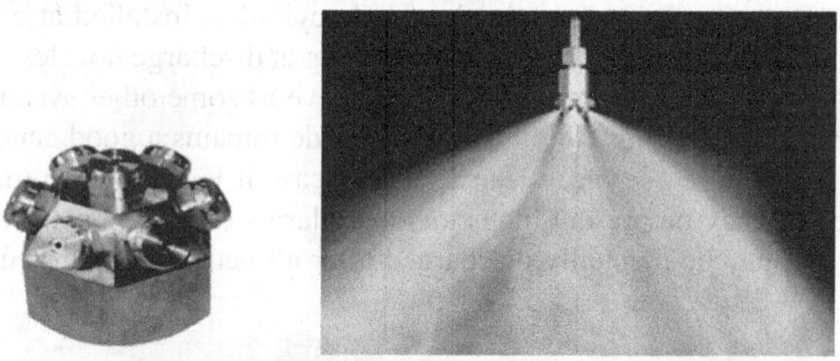

Figure 9.16 Water mist nozzle and discharge.

- provide a large surface area for cooling,
- vaporise rapidly into steam which helps displace oxygen, and
- form a cloud to attenuate radiation feedback.
- cool combustion gases.

The smaller a water droplet size is, the larger the surface area due to the larger number of water globules and heat removal is more effective by water converting to steam. It should not, however, be assumed that smaller globule size always means better because the globules need momentum to enter the plume or otherwise drop downwards and not be blown away.

Water mist protection can be designed for use to

- protect critical single items such as generator sets or deep fat fryers in commercial kitchens and
- flooding to protect volumes such as control equipment spaces, data processing centres or cabins on ships.

Water supply to water mist systems can be either through pumps or through pressurised gas systems. The discharge nozzles may be open or closed types where the latter is linked to a detection system for automatic actuation. Water mist systems may be designed to be deluge or individually acting nozzle type. The working pressures may also vary depending on system intent and designed accordingly. With such a wide range of water mist systems available, it is essential that management of the premises conduct a detailed risk assessment to identify their needs before an expensive, specifically designed and carefully maintained system is selected and installed.

9.7 CARBON DIOXIDE FLOODING SYSTEMS

Carbon dioxide has for long been recognised as an effective fire fighting medium and has been used in extinguishers of different sizes as also in fixed protection systems. A primary concern in the use of carbon dioxide is life safety as the gas extinguishes fire by displacement of air from the fire area. This can be dangerous in enclosed spaces such as small rooms. Further, fixed carbon dioxide systems are required to discharge large volumes of the gas to achieve extinguishment by volume filling. The discharge of the gas both from extinguishers or fixed systems also generates loud noise which may unnerve people. This though is indicative of the gas being discharged.

Carbon dioxide systems comprise a bank of CO_2 cylinders installed at a safe location with pipework leading into the protected area and terminating at discharge nozzles spread out to reach all sections (Figure 9.17). The system is not as expensive as some other systems like water mist systems and with the phase-out of Halon, carbon dioxide remains a good candidate for flooding unmanned areas that require use of a clean agent. In case it is decided to install such systems in areas where people may be present, detectors and alarms to alert people to evacuate may be installed and the gas may be manually discharged after all people have evacuated.

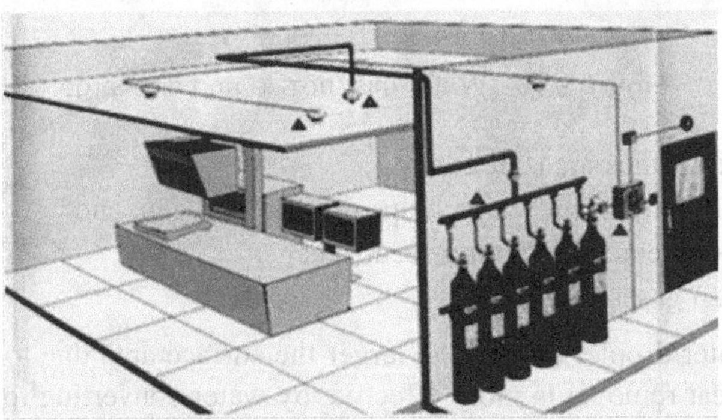

Figure 9.17 Carbon dioxide protection system.

9.8 CLEAN AGENTS

The National Fire Protection Association defines, a *Clean Agent* as an electrically non-conducting, volatile, or gaseous fire extinguishant that does not leave a residue upon evaporation.

The very popular gaseous fire and explosion suppression agents of the 1970s and 1980s viz., Halons (halogenated hydrocarbons) needed to be phased out (with few exceptions such as civil aviation) due to the stipulations of the Montreal Protocol regarding ozone depleting substance (ODS). Halons leave no damaging residue and can be used in required discharge concentrations even when people were present.

This necessitated a search for 'Halon alternatives', that is, clean agents that could be used in lieu of the much preferred Halons. In response, extinguishing media based on hydrofluorocarbons (HFCs), perfluorocarbons (PFCs), chlorofluorocarbons (CFCs) and hydrochlorofluorocarbons (HCFCs) as well as inert gas have been developed. Some of these are HCFC-123, HCFC-22 and HCFC-124 agent blends.

For fixed systems, available alternatives are carbon dioxide, inert gases, HFCs, PFCs, HCFCs and fluoroketone (FK). The factors that contribute to their selection are performance, life safety, cost and environmental concerns. Extinguishers using HFCs, PFCs and HCFCs have, however, have not found popularity, mainly due to their high cost. However, the cost of manufacturing these products remains high and their use has not become very popular.

9.9 DRAINING OUT FIRE WATER

Water-based fire protection systems such as cooling water systems, foam/water monitors, hose streams, etc. together use vast quantities of water and there must be adequate provision to drain out the used water after treatment where that may be required. During an actual fire scenario, it may not also be possible to apply water at 'design' rates and excess water may be applied. There are landmark instances when indiscriminate release of fire fighting water has polluted long stretches of rivers. On other instances, tank dykes have got flooded due to inadequate provision for drainage and led to 'floating' of nearly empty tanks.

Thus, it is important to size dyke drains for various situations including massive fire fighting operations. Such releases must also pass through effluent treatment plants to prevent escape of polluting streams.

QUESTIONS FOR REVISION

1. Name and explain the purpose of the primary components of a hydrant system.
2. What are the differences in purpose of automatic sprinklers and medium-velocity water spray systems? Can one be replaced by the other?
3. What are the differences in purpose of high-velocity water spray systems and medium velocity water spray systems? Can one be replaced by the other?
4. What are the important characteristics of a fire fighting foam? How are they important in effective fire fighting?

5. What would be the effect of using a foam making branch meant for 3% foam compound on 6% foam compound? What would be the effect if the reverse is done?

6. What type of foam would be appropriate for use in petrol with ethanol blend? What would be the effect of using foam meant for non-polar fuels on such a blend?

7. Give examples of locations where high-expansion fire fighting foams would be the appropriate medium to be used.

8. What are the precautions to be taken from life-safety point of view in case of fixed fire protection systems based on carbon dioxide?

9. Study about the various clean agents being developed as replacements for Halon. Where is Halon still permitted to be used despite its ozone depleting property?

10. Why is emulsification by high-velocity water spray systems not recommended for use on fires in light hydrocarbons?

10

Risk Management and Fire Insurance

What has not been identified cannot be evaluated and hence cannot be treated.

—Anon

10.1 ORIGINS OF GENERAL INSURANCE AND EMERGENCE OF RISK MANAGEMENT

Insurance in different forms is reported to have existed as early as in the Babylonian times but, insurance as we know today got shape in the seventeenth century at Lloyds coffee house in London where seafarers met to put aside sums of money to mitigate losses in transit experienced by their unfortunate peers. Thus was born modern marine insurance. Around the same time, following the 1666 Great Fire of London, England's first fire insurance company, 'The Fire Office' was established to provide insurance cover against fire loss events. Some decades later, in the mid-eighteenth century, Benjamin Franklin set up his fire insurance company in USA. Franklin's company refused to insure buildings where the risk of fire was very high (as in timber structures) and thus encouraged risk improvement practices.

Reference to insurance is found in ancient Indian texts like Manusmriti, Dharmashastra and Arthashastra, which refer to pooling of resources that could be redistributed in times of calamities such as fire, floods, epidemics and famine. Modern insurance in India has evolved over time drawing largely from practices abroad and particularly from Britain.

In India, life insurance business in its modern form was started by the Oriental Life Insurance Company in Kolkata in 1818, while the Triton Insurance Company Ltd. started operations as the first general insurance company in 1850, also in Kolkata. The earliest effort at regulating insurance business in India was through the Insurance Act, 1938. Thereafter, many changes have taken place both in life insurance as well as in general insurance business. Effective 31.03.2000, fire insurance business is regulated through premium ratings done by the Tariff Advisory Committee (TAC), a statutory body under the Insurance Act, 1938. Terms of general insurance business in India are determined by the Insurance Regulatory and Development Authority (IRDA) set up under the IRDA Act, 1999.

In due course, people realised that money received against claims could not compensate all losses and that it was advisable to take preventive and protective actions. Simultaneously, insurers also adopted the concept of reward and punishment by which properties that were better protected were offered discounts in premiums while those with less protection were penalised through higher (loaded) premiums. The combined effect of these practices possibly led to the development of the field of ***risk management,*** wherein insurance occupies the last stage viz., risk transfer. This chapter will discuss on the stages of risk treatment starting from risk avoidance through emergency management with the intermediate stages of risk reduction and insurance.

10.2 HAZARD VERSUS RISK

Terms like risk, chance, hazard, likelihood and probability are commonly used very loosely and often as synonyms. For instance, the statement that one is taking a risk could mean that one is taking a chance; referring to an activity as being risky could mean that it could be hazardous. Let us, therefore, try and understand these terms before going on to discuss salient aspects of the process of risk management.

Hazard implies the characteristic of a system or any of its components with potential to cause a loss of any resource. Thus, pyrophoric iron, high-temperature processes, flammable vapour–air mixtures, pressurised storage and bad housekeeping are examples of fire hazards. It only requires a triggering action for a hazard to cause loss. In other words, a fire hazard may be the reason for a fire to start when other conditions that are required for the fire process are satisfied.

Likelihood ↑		
	High likelihood, low consequence	High likelihood, high consequence
	Low likelihood, low consequence	Low likelihood, high consequence
	Consequence ⟶	

Figure 10.1 Likelihood versus consequence diagram.

Risk is a combination of the likelihood that a hazard could give rise to a loss and the potential severity of the loss (Figure 10.1). Likelihood indicates the probability of occurrence of a loss event. Severity may be measured in terms of loss of life or scale of personal injury, loss of or damage to assets, the envelope or boundaries of damage, extent of damage to the environment or the loss of reputation for the concerned entity. Thus, risk is a function of likelihood and severity, or,

$$\text{Risk} = f\,(\text{likelihood, severity})$$

Thus, *fire risk assessment* involves assessing how likely is it that a fire will occur in a given system and, if it does occur, how damaging its results could be. This is an essential first step to identify the methods to adopt to reduce the likelihood of occurrence of a potential fire incident and thereafter control the adverse effects of the incident. It is important that both likelihood and severity be considered to assess any risk — if only severity is considered, assessment thereof is termed *consequence assessment*.

To recapitulate, the first step in risk management is hazard identification through which efforts are made to identify all hazards of an activity, process or location. Next, hazards need to be analysed to recognise routes to loss events. Hazard identification and analysis form the basis of risk assessment which involves evaluation of the likelihood that a loss event could occur and the quantification of the potential consequence of the loss event.

Interestingly, ***likelihood*** estimates are often made on the basis of ***frequency*** though the former refers to future events, while the latter is based on history or past experience and system improvements could have been introduced at different stages.

10.3 HAZARD IDENTIFICATION AND ANALYSIS

Hazards associated in an activity may not always be obvious and specialists may be required to identify hazards and potential loss scenarios in complex and capital intensive processes and facilities. Whether it is to design a safe system or facility, to root out hidden hazards in the workplace or to have appropriate emergency preparedness plans in place, it is essential that hazards are identified and analysed and their risks assessed. There are different tools and techniques that can be used for hazard identification but the effort required, applicability and the outputs are different. An understanding of the different techniques is, therefore, required to choose the appropriate one.

10.3.1 Checklists

Standards and statutes have been used to prepare *checklists* for use by designers, risk engineers and managements. Some industries have also developed their own checklists suited to their facilities and activities. Checklists are easy-to-use hazard identification tools particularly if resources like time and expertise are unavailable or if the occupancy is assessed as not requiring involved exercise. They are commonly used in existing facilities to verify compliance and deviations against standards (audits or inspections) (Figure 10.2). The Oil Industry Safety Directorate (OISD) has developed detailed safety audit checklists for different sectors of the hydrocarbon industry. Checklists may, however, be inadequate in case of new technology, obsolete or generic stipulations, co-existence of dissimilar activities, etc. Checklists are prescriptive in nature and cannot indicate either the cause or the potential outcome of a deviation, or appropriate preventive and protective actions. Thus, while checklists are useful tools, the scope of a hazard identification exercise should be clear to determine whether a checklist is indeed appropriate.

10.3.2 Preliminary Hazard Analysis

As the name indicates, preliminary hazard analysis (PHA) is intended to help in conducting preliminary analyses of hazards in an activity or process. Both the inputs required and the results arrived at are not detailed. On the other hand, PHA offers a quick, qualitative assessment as also means of rectification Figure 10.3. Shows a typical PHA format. Columns may be added to add details like route to hazardous events, severity of effects, etc.

Location:		Date:	
Description of occupancy and type of construction:			
Team members: 1.		2.	
Part 1: Fire safety features			
Sl.No.		Yes	No
1.	Is the premises connected to other premises?		
2.	Are exit routes marked and kept free?		
3.	Are aisles marked and kept free?		
4.	Are extinguishers provided as per IS:2190?		
5.	Are extinguishers maintained as required and records kept?		
6.	Are automatic detector and/or alarm systems installed on site?		
7.	Are detectors/alarms linked to a central panel monitored 24×7?		
8.	Are internal hose reels provided and maintained?		
9.	Is the facility provided with hydrant protection?		
10.	Are personnel trained in fire fighting?		
11.	Are mock drills conducted on site?		
12.	Are external fire escape stairs provided in multi-storeyed buildings?		
13.	Are personnel aware of emergency procedures?		
14.	Is a permit-to-work system followed for hot work?		
15.	Are fire safety audits/inspections conducted regularly?		
16.	Are MSDS available on site?		
17.	Are hazardous materials marked and stored properly?		
18.	Are fire incidents investigated and records maintained?		
19.	Does the site have access for emergency vehicles?		
20.	Are statutory licences available and up-to-date?		

Figure 10.2 Sample fire safety checklist.

Activity	*Hazards*	*Potential effects*	*Corrective and preventive measures*

Figure 10.3 Format for PHA.

10.3.3 Logic Diagrams

Logic diagrams such as *fault trees, event trees* and *cause-consequence diagrams* are designed to use logic to reason out sequence of events leading to undesired events. Fault tree analysis

(FTA) involves finding out potential causes (i.e., hazards) that may have caused a fault or failure event and hence is termed *backward looking*. Event tree analysis (ETA), on the other hand, aims to identify potential outcomes from a given scenario and is termed *forward looking*. Cause-consequence analysis (CCA) is a combination of FTA and ETA designed to look for causes leading to an event as well as potential outcomes from the event. Logic diagrams can be used for qualitative analysis and, if data is available, they can be used for quantitative analysis as well. Some simple examples of FTA (Figure 10.4) and ETA(Figure 10.5) are illustrated below.

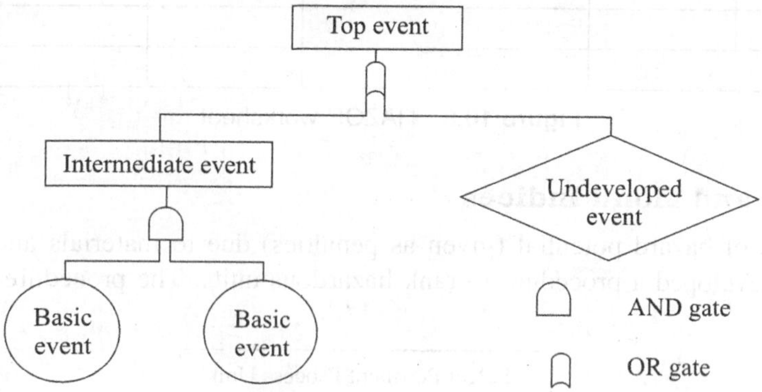

Figure 10.4 Fault tree.

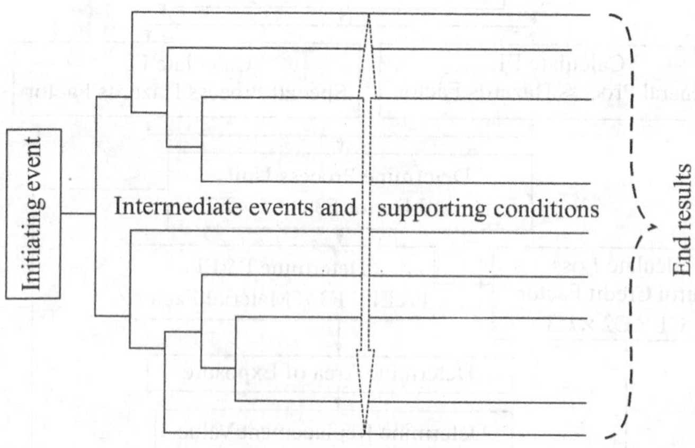

Figure 10.5 Event tree.

10.3.4 Hazard and Operability Study (HAZOP)

The expanded form of HAZOP is ***hazard and operability study*** and is a risk assessment tool based on the premise that undesired incidents occur due to deviations of set control parameters (Figure 10.6). Initially developed by and for the chemical industry involving material flow and process, the tool has been adopted for use in other industries as well as evidenced by the international standard IEC 61882–2001, Hazard and operability studies (HAZOP studies) — Application guide. Guidewords indicating deviations of parameters are used to arrive at possible causes thereof and their potential results.

Guideword	*Parameter*	*Cause*	*Effect*	*Corrective action*
No				
More	Flow	Failure of flow controller in open position		
Part of				
Less of				
As well as				
Other than				
Reverse				

Figure 10.6 HAZOP worksheet.

10.3.5 Dow and Mond Indices

Based on factors of hazard potential (given as penalties) due to materials and processes used in a unit, Dow developed a procedure to rank hazardous units. The procedure helped to arrive

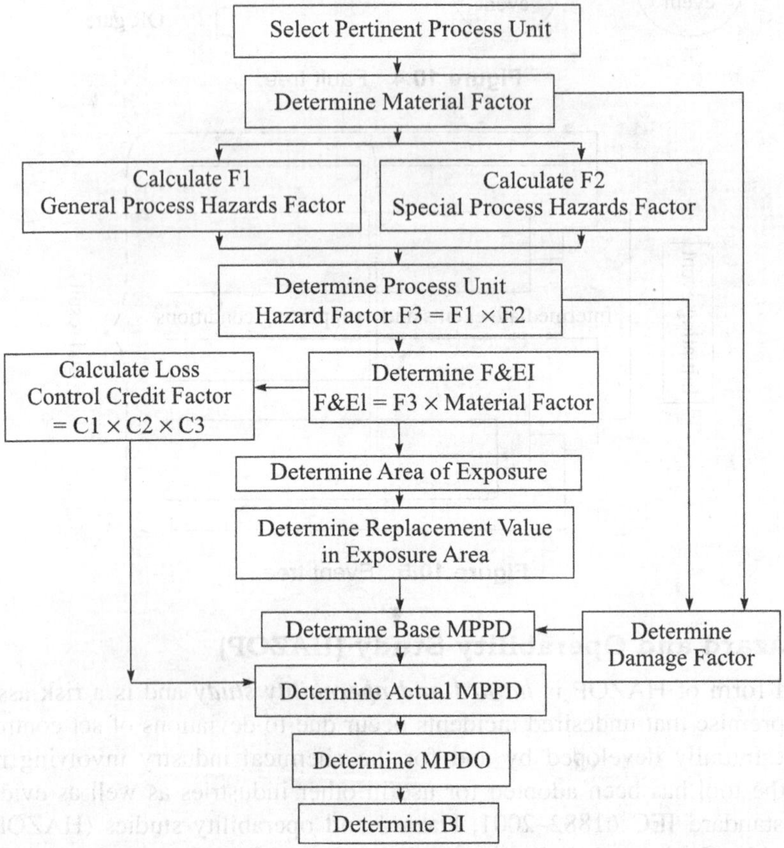

Figure 10.7 Procedure for calculating Dow Fire and Explosion Index.

at a numerical value known as the Fire and Explosion Index (F&EI) and has been revised and upgraded several times. At one stage, the benefit due to protective measures (credit factors) was introduced as opposed to the 'penalties only' concept in earlier versions. Later the Mond unit of the erstwhile ICI group modified the procedure to include the aspect of toxicity as well as enable the tool to address processes with explosive potential. The index thus arrived came to be known as *Mond Index*.

The F&EI can be used to address the area of exposure and thus the maximum probable property damage (MPPD), maximum probable days outage (MPDO) and business interruption (BI) (Figure 10.7). These are useful data used by insurance engineers in their insurance rating exercises.

10.3.6 Computer Models

The large number of variables involved in the occurrence of outcomes like fires or explosions often make it almost impossible to assess risks manually. Computer modelling has to a great extent provided the answer to address this difficulty and today there is a large amount of work being done in this area. Models have been developed for compartment fire scenarios, smoke movement in enclosed spaces, evacuation procedures, formation and dispersion of vapour clouds, etc., all of which have provided vastly improved outcome prediction capabilities.

10.3.7 Others

There are several other tools for hazard identification that have been developed and used, but there is no single tool that may be used effectively in all cases and applicability of different tools will depend on the nature of activities, resources available vis-à-vis resources required, the type of decision making that is called for, etc. The results obtained by using the various tools also vary widely.

10.4 RISK ASSESSMENT AND CONTROL

After hazards have been identified and analysed, the risk thereof viz., the likelihood to cause loss and the potential consequences need to be assessed. This activity, known as *risk assessment*, may include the following steps:

- Assessing the likelihood of the loss event.
- Assessing the potential consequences.
- Describing how it could become a liability to the organisation.

Based on the outcome of this assessment, various options of risk treatment can be considered and the most appropriate selected. The following sequence of risk treatment is generally adopted:

- Risk avoidance or elimination
- Risk reduction
- Risk retention and
- Risk transfer.

If risk from an activity is assessed as 'high', the first option is ***risk avoidance*** or ***risk elimination***. This may imply either avoiding a business activity or outsourcing it if that is feasible. In case risk elimination through these routes is not feasible, one has to move to the step of risk reduction. An interesting approach to a high-risk situation is voluntary acceptance as one does in a dangerous sport like mountaineering. In such instances, we may use the term ***acceptable risk***; otherwise the appropriate term to be used is ***tolerable risk***.

Risk reduction options can include decisions like using less hazardous processes or materials, reducing material inventories, isolating hazards through segregation, installing trips or damage control provisions like automatic sprinklers, adopting safe work procedures and developing emergency management plans to address realistic loss scenarios. Means selected for risk reduction should be feasible, viable and appropriate for the identified hazards. It is well-established that risk cannot be eliminated and reduction should be exercised till ALARP (as low as reasonably practicable) level is reached.

Residual risk remaining after the risk has been reduced to ALARP should be estimated to determine means of mitigation. If in-house resources can adequately address the risk, the risk may be retained and the decision is called ***risk retention***. If residual risk is estimated as high, some or all of it may be transferred (***risk transfer***) to another party. This is typically done through a contract of insurance.

In the foregoing discussion, we have used the term *high risk*. However, what comprises high risk can be a subjective assessment and could involve trade-offs of different types. Constraints such as availability of resources for risk reduction and the level of risk that may be retained can also play a role in decision making. Hence, while different options of risk management should be considered in all cases, it would be wrong to offer standard menus that are appropriate for universal use.

It is important to remember at all times that risk has two components of likelihood and severity and risk assessment should consider both of them. Analysts may be swayed by severity alone while likelihood is ignored. This could result in expending resources on major impact events that are very unlikely to occur. Good risk management plans evolve by experience and professional judgement. In case of complex processes or facilities where in-house expertise may not be available, help of expert consultants may be sought to draw up cost-effective risk management plans but with the participation of in-house representatives.

10.5 PRINCIPLES OF INSURANCE

Comparing insurance with gambling or speculation is inappropriate as the natures of uncertainties in the two are very different. In speculation, as in the case of stock market dealings or laying bets, there is scope for both gain as well as loss. Insurance, on the other hand, addresses only losses that could come about due to uncertainties; there is no scope for gain or profit. Thus, insurance is not speculation and deals with the subject of 'pure' risk while gambling or speculation deals with 'speculative' risk. Insurance is intended to deal with uncertainties, the exception being life insurance where end is certain. In life insurance, uncertainty is the time of death and that is where insurance comes in.

In a contract of insurance, the two parties are the ***insurer*** and the ***insured***. The principles of insurance require that when proposing an insurance contract, both the 'proposer' (after the contract is signed, he/she becomes the insured) and the insurer must observe the following principles:

- Insurable interest
- Utmost good faith
- Indemnity

Insurable interest means that the insurer must be an owner, a trustee or otherwise have an interest in the assets or activity being insured. Thus, one cannot seek insurance cover for material or property that either does not belong to him/her or damage/loss to which would not in any way cause him/her fiduciary loss.

Utmost good faith requires that the proposer (insured after signing of the contract) does not hide from the insurer any material facts about the assets or activities being insured.

Indemnity implies that an insurance cover can only serve to make good the insured losses and cannot be used to earn profit for the insured.

10.6 FIRE INSURANCE POLICY

From 31.3.2001, fire insurance in India is governed by All India Fire Tariff issued by Tariff Advisory Committee, a statutory body under the Insurance Act, 1938 and under the administrative control of IRDA.

Some highlights of the Indian Standard Fire and Special Perils Policy are:

- Covers all moveable or immoveable properties on land against named perils.
- Additions/deletions of perils with additional/discounted premium possible.
- Normally issued for a year. Long-term policies with discounts in premium available for dwellings.
- Discount in premium available for good claims experience in special cases.

10.7 COVERS AVAILABLE

10.7.1 Perils and Expenses

The Standard Fire and Special Perils Policy provides cover against destruction of or damage to insured property due to incidental fire as also some other causes, known as ***perils*** in insurance terms. They are:

- Fire, explosion/implosion, lightning.
- Impact by any rail/road vehicle, animal or aircraft damage.
- Riot, strike, malicious and terrorism damage (RSMTD).
- Storm, tempest, flood, inundation, etc. (Also known as STFI perils)
- Subsidence/landslide including rockslide.
- Bursting and/or overflowing of water tanks and apparatus.

- Leakage form automatic sprinkler installation.
- Pollution or contamination from any of the above perils or the reverse.
- Bush fire.
- Missile testing operation

In addition, the policy also covers some ***expenses*** incurred (up to a limit) after a loss event such as fees of surveyors and expenses for removal of debris.

Extensions available include loss/damage/destruction of the property caused by perils such as

- Earthquake (fire and shock).
- Spontaneous combustion.
- Deterioration of stocks in cold storage due to power failure following damage due to an insured peril.
- Forest fire.
- Impact damage due to insured's own vehicles, forklifts and the like and articles dropped from them.
- Omission to insure additions, alterations or extensions.
- Spoilage material damage cover.
- Leakage and contamination cover.
- Temporary removal of stocks.

Extensions available for expenses include start-up expenses, loss of rent, rent for alternate accommodation, excess professional fees and excess cost of removing debris.

10.7.2 Properties

The policy covers all moveable/immoveable properties of the insured on land (excluding those in transit) and broadly categorised as follows:

- Building.
- Plant and machinery, equipment and accessories.
- Stocks including raw material, finished goods, in process or in trade.
- Other contents such as furniture and fixtures, spares, tools and stores, household goods, etc.
- Specific items such as bullion, unset precious stones, works of art, plans, drawings, manuscripts, cash, securities, cheques, books of accounts, computer records and explosives.

10.8 INSURANCE VALUE

Amounts recoverable through an insurance policy after a loss event are often contested and hence a listing of ways of valuation may be relevant.

- **Buildings:** If completed, based on either depreciated value or reinstatement value; if under construction, reinstatement value.
- **Plant and machinery:** If bought new, market value or reinstatement value; if bought second, market value; if obsolete, agreed value ascertained by an expert valuer.

- **Stocks:** For raw material, finished goods, stock in process, market value.
- **Stock-in-trade:** For manufacturer/wholesaler/retailer, market value.
- **Furniture, Fixture and Fittings:** Market value or reinstatement value.
- **Items covered specifically on declaration:** Agreed value as certified by an expert valuer.

10.9 REINSURANCE

Although it may appear as only of academic interest to insureds, one may wonder about the resources of an insurance company to pay for very large losses. To cater for such eventualities, insurers buy reinsurance, which may be described as insurance for insurers. It is an agreement between an insurer (cedent) and a reinsurer wherein the reinsurer agrees to indemnify the cedent against all or part of a loss, which the cedent may incur from a policy it has issued. For this, the cedent pays a consideration in the form of premium to the reinsurer.

There are large and small reinsurance companies worldwide that offer reinsurance. For Indian insurers, however, the IRDA Act, 1999 and the Insurance Act, 1938 govern and regulate reinsurance practice. In November 2000, General Insurance Corporation of India (GIC) was renotified as the Indian reinsurer and GICRe was born. This entitles it to receive obligatory cessions of 20% from all the non-life insurers operating in India. If Indian insurers choose to place their reinsurance business outside India, they may do so only with re-insurers approved by IRDA. Later, when the General Insurance Business (Nationalisation) Amendment Act 2002 came into force on March 21, 2003 GIC ceased to be a holding company of its subsidiaries and their ownership was passed on to the Government of India.

10.10 EMERGENCY PREPAREDNESS

Industry managements realise that notwithstanding good design, erection, operation and maintenance, the likelihood of emergencies cannot be eliminated. And yet, when the need to tackle an actual emergency does arise, often the preparedness proves inadequate or inappropriate. Insurance cannot and does not cater to all losses. In fact, as mentioned earlier, a condition of an insurance contract is that, in case of an incident, "the insured shall act as if not insured" implying that loss control actions must be undertaken during an incident like calling the fire brigade and initiating fire fighting actions. It is in this regard too that appropriate emergency plans to cater to emergency scenarios must be in place and they should be activated if and when an emergency does occur.

Emergencies due to a business activity can originate either within a manufacturing, handling or storage facility or in external activities like transportation. Within the former category, the effects can be confined to either inside the facility or extend beyond it. When an emergency originates inside the facility and its effects are confined to its boundary, the scenario is commonly called an *on-site emergency*. To deal with such an event, the control and responsibility lie with the facility management although they may seek the help of public services like the fire brigade. Pre-planning for effective handling of the scenario is done through an ***on-site emergency plan***. When the effects extend beyond the boundary of the facility or, if the emergency originates outside, the scenario is termed *off-site emergency* and it has to be dealt through an ***off-site emergency plan***. As the affected area in this case is likely to be a public area, facility management may have no jurisdiction and the concerned public authority has to take charge of the situation.

Following the Union Carbide, Bhopal disaster (1984), there were major amendments made in 1987 to the Factories Act, 1948 primarily to address disasters related to chemicals. Thus was born the Control of Major Accident Hazard (CIMAH) Rules promulgated by the states under provision of Section 41B of the Act. Further, the Manufacture, Storage and Import of Hazardous Chemicals (MSIHC) Rules, 1989 and the Chemical Accidents (Emergency Planning, Preparedness and Response) Rules, 1996 both framed under the Environment (Protection) Act, 1986, it is mandatory for hazardous installations to institute plans to handle different 'on-site' emergencies including fires and explosions.

The Chemical Accidents (Emergency Planning, Preparedness and Response) Rules, 1996 stipulates the formation and roles of crisis management groups at the central, state, district and local levels.

With the passing of the Disaster Management Act, 2005 and the establishment of the National Disaster Management Authority (NDMA) as laid down in the Act, the subject of disaster management has been given a better structure. The Act addresses both 'natural' as well as 'man-made' disasters.

The MSIHC Rules, 1989 and the Chemical Accidents (Emergency, Planning Preparedness and Response) Rules, 1996 require industries to have their own emergency plans. In addition, industries are also required to assist the local administration draw up 'off site' emergency plans.

In the case of commercial buildings, hospitals, residential blocks, educational institutions, places of entertainment, however, there are no statutory provisions in India that require emergency management, evacuation or rescue plans. This should not be construed as reason not to have emergency plans and managements must conscientiously draw up appropriate plans to contain and control the effects of anticipated emergencies.

10.11 FEATURES OF EMERGENCY PLANS

10.11.1 Basis of Emergency Plans

An emergency management plan should begin with identifying and assessing all potential emergency scenarios that may arise from activities in the facility and their routes of likely development with stages and time scale till they may emerge as off-site emergencies. While risk assessment essentially involves estimating both likelihood and consequence, it may be difficult to arrive at numerical values of likelihood in many cases and managements may have to make their own estimates of likelihood which may be qualitative. Remotely potential emergencies like an aircraft crash on the facility, however, may be excluded.

Emergency plans have to be realistic and designed to address scenarios that are likely to occur in a facility as identified using reliable techniques. The components of an emergency plan may be generic but details must be drawn out for a specific location and its activities. Emergency plans must take into account the following factors:

- Likelihood of an incident.
- Potential consequences of the incident on people and assets (within the facility and outside) and on the environment.
- Resources required for combating the emergency and resuming business.
- Resources available in-house and those to be arranged from other sources.

10.11.2 Estimating Resource Requirements

In the next stage, resources that may be required to control and contain the scenario are assessed and gaps identified against resources available in-house or within easy access. While metros and industry centres may have adequate inventories of resources, locations where public emergency services may be minimal need to upgrade their own emergency handling provisions. Requirement and deployment of these resources have to be planned for different scenarios. This is a good time to discuss the plan in detail with the local administration both in terms of potential off-site emergency scenarios and emergency management resources.

10.11.3 Emergency Management Teams

An emergency organisation has to be established with roles and responsibilities defined, procedures made and widely publicised. In-house organisational teams may have the following team members selected appropriately for their authority, skills, organisational role, etc.:

- Incident controller
- Resource coordinator
- Media coordinator
- Communications centre/coordinator
- Transport/welfare
- Utilities coordinator
- Fire and rescue chief

An off-site emergency inherently implies that the effects extend to areas beyond the jurisdiction of the facility management. Hence, the public authority having jurisdiction necessarily has to take charge of the situation and public agencies are essentially active players in the execution of an off-site emergency plan. Some public agencies that may be required to play a role in tackling an off-site emergency are:

- Civil administration
- Police
- Fire brigade
- Pollution Control Board
- Transport authority
- Health services
- Veterinary department
- Fisheries department
- Forest department

10.11.4 Drills

While tabletop drills are commonly used to practice role play and check efficacy of emergency management plans, its use in verifying different aspects of an off-site plan is doubtful. Hence, off-site emergency plans have to be rehearsed at site for various possible scenarios with observers to chronicle all facets of the drills. After a rehearsal, a debriefing session is essential to identify lacunae and recommend improvements.

10.12 LINE OF ACTION

Figures 10.8 and 10.9 represent stages of potential emergency scenarios, their likely effects and the basis for developing emergency plans to address these scenarios.

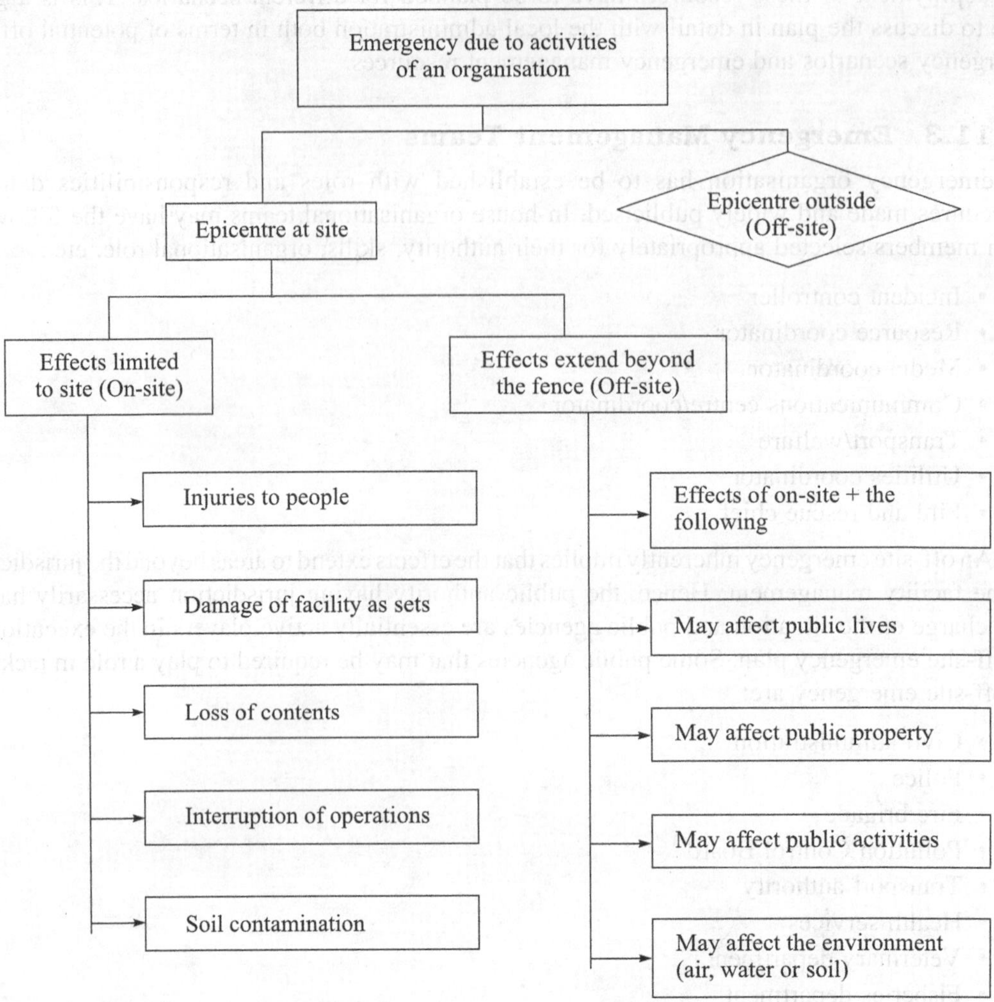

Figure 10.8 Potential effects of emergencies.

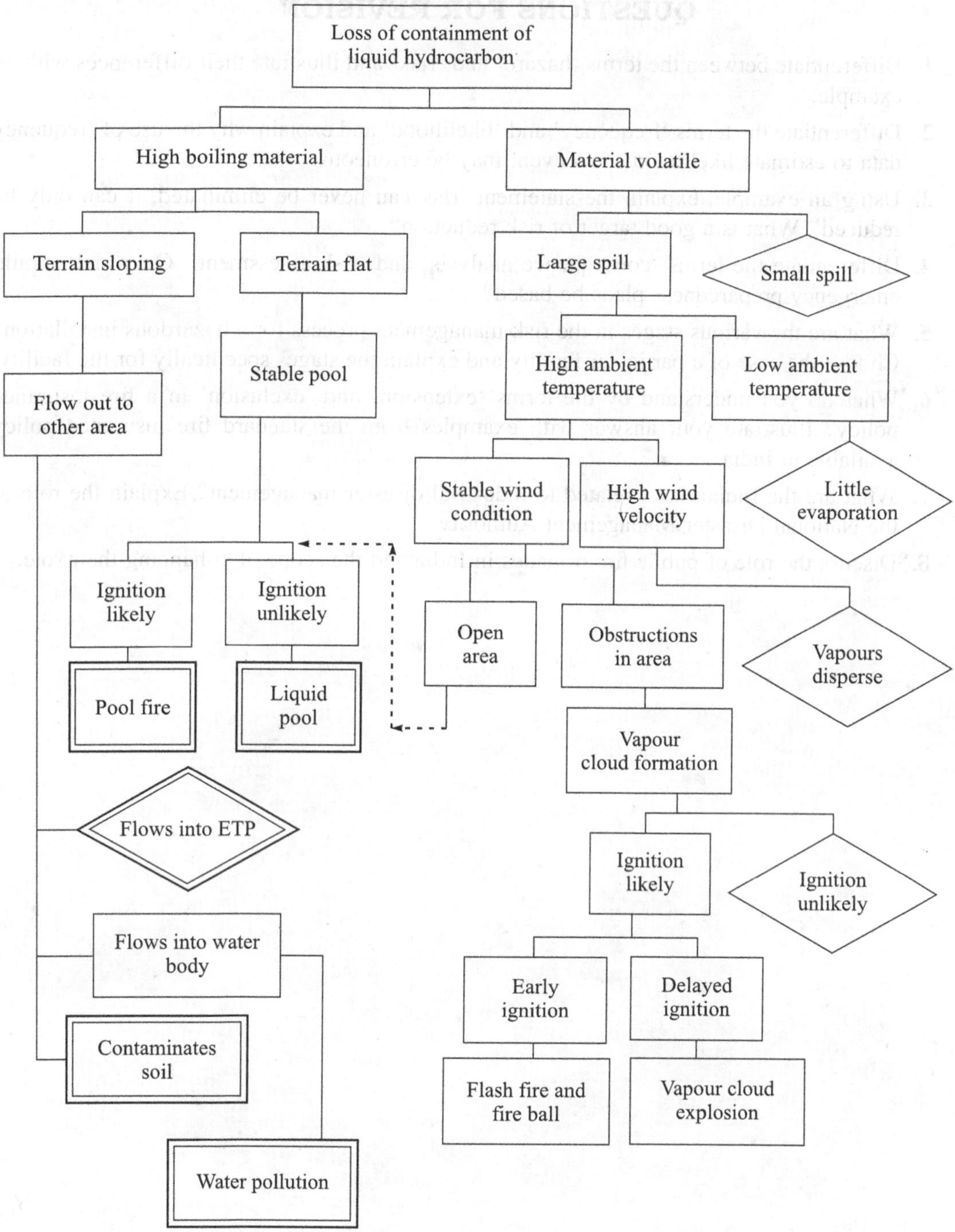

Figure 10.9 Event sequence diagram for a hypothetical loss of containment of flammable liquid.

QUESTIONS FOR REVISION

1. Differentiate between the terms 'hazard' and 'risk' and illustrate their differences with an example.
2. Differentiate the terms 'frequency' and 'likelihood' and explain why the use of frequency data to estimate likelihood of an event may be erroneous.
3. Using an example, explain the statement 'risk can never be eliminated, it can only be reduced'. What is a good target of risk reduction?
4. Differentiate the terms 'consequence analysis' and 'risk assessment'. On which should emergency preparedness plans be based?
5. What are the various stages in the risk management process for a hazardous installation? Choose the case of a particular facility and explain the stages specifically for the facility.
6. What do you understand by the terms 'extension' and 'exclusion' in a fire insurance policy? Illustrate your answer with examples from the standard fire insurance policy available in India.
7. What are the Indian laws related to industrial disaster management? Explain the role of the National Disaster Management Authority.
8. Discuss the role of public fire brigades in India and the scope of enhancing their role.

11

Investigating Fire Incidents

"It is a capital mistake to theorize before one has data. Insensibly one begins to twist facts to suit theories instead of theories to suit facts."

— Sherlock Holmes in "A Scandal in Bohemia"

11.1 BACKGROUND TO FIRE INVESTIGATION

People often tend to look at fire incidents as instances either of misfortune or of heavenly wrath. However, not all fires are 'acts of God' and all fires need to be investigated for legal, insurance and social reasons or to learn lessons and develop better understanding of the fire process. Statistics generally categorise fires based on the 'likely' source of ignition, which is also referred to as 'cause' and 'electrical short circuit' is a frequently named candidate. Such immediate attribution of fire causes indicates improper investigation. Fire investigation should aim to identify probable origin and cause of a fire by gathering evidence and trying to reconstruct the pre-fire scene as well as the sequence of events with the evidence collected which includes examining damaged assets and sieving through debris to obtain physical evidence and interviewing occupants, users or informed people. Thorough investigation also helps fire professionals improve their knowledge of the fire process and its effects and thus enhance methods of fire protection. Another purpose of fire investigation is to identify any malicious intent (commonly known as arson) that may arise out of reasons such as revenge, sabotage, vandalism and fraud.

Fires may be categorised on the basis of their origin viz., natural, accidental or arson:

- ***Natural*** fires may start from natural causes like lightning.
- ***Accidental*** fires may result from human activity gone out of control.
- ***Arson*** fires started deliberately with malicious intent.

Irrespective of the type, major fires combined with associated fire fighting activities obliterate most footprints that could help investigators reconstruct a pre-fire scenario. The activity of fire investigation thus is not a layman's field and it requires specialist investigators to mine out evidence in the midst of debris and destruction. Observations made by fire fighters and other eye witnesses, debris analysis by forensic scientists, study of previous records and interpretation by scientists and engineers — all are areas that need to be addressed for a comprehensive fire investigation. This chapter tries to highlight some salient facets of fire investigation and is not in any way a comprehensive guide to fire investigation.

11.2 PRACTICE OF INVESTIGATION

Fires in private or public buildings pose very different issues. Except when a property is insured or if arson is suspected, single home fires may be only superficially investigated. An apartment building fire may receive attention if it involves more than one dwelling. Fires in educational premises can initially attract much public and media attention but, unless there is loss of life, investigations into fires in schools, colleges or universities are not known to be thorough. Fires in public buildings like those housing offices, places of entertainment, hotels, hospitals and temporary structures like pandals may affect public life and property and are widely reported but detailed investigation is not normal.

The objectives of a fire investigation could be many but, in all cases, they should include, among other things:

- The place where the fire started (origin or seat of fire).
- The likely source of ignition.
- The factors that led to fire growth.
- Quantum of loss to assets and other resources.
- Injuries or fatality (if any).
- Efficacy of fire detection and control systems.
- Fire fighting efforts and resources used.
- The type of fire, viz., natural, accidental or deliberate.

It should be evident from the above that fire investigation is a multi-faceted activity requiring a variety of skills to unearth likely processes, materials and activities to explain the outcome of the fire.

11.3 ARSON

Arson has existed even in the times of epics. Arsonists are known to have started fires for profit such as through insurance claims, to hide some other crime, in revenge, sabotage, even for fun (pyromania) or many other reasons. If arson is a likely cause of fire, the investigator has to ascertain if it indeed was so.

To ensure success of their intent and accelerate the burning process, arsonists may use flammable liquids as ***accelerants***. Thus, when arson is suspected, it is important to search for, locate, sample and analyse debris for traces of accelerants. This will require help of forensic analysis. Arsonists may also use other means such as explosives, pyrophoric substances, fire crackers, creating electrical faults to ensure completeness of burning.

11.4 WHO SHOULD INVESTIGATE FIRES?

The National Fire Protection Association, USA has two standards dedicated to Fire Investigation (2011 NFPA 921 Guide for Fire and Explosion Investigations and 2014 NFPA 1033 Standard for Professional Qualifications for Fire Investigator). The titles of the two standards are indicative of their contents. The former provides guidelines and recommendations for safe and systematic investigation of fires and explosions and the latter signifies the importance of 'qualifications' to be a fire investigator. Specialist professional agencies like the Chemical Safety and Hazard Investigation Board (CSB) and the National Transportation Safety Board (NTSB) — both also of the US — also conduct investigations in their respective areas of concern.

In India, two types of fire investigations are common:

- Fires in industry are investigated by technical specialists from different disciplines and
- Fires in the public domain are generally investigated by public fire brigades unless courts or government agencies specifically institute specialist investigation.

In the former, investigation is generally done by multi-disciplinary teams and is based on trying to identify modes of failure of material, equipment or process. Investigation by public fire brigades may be done by the attending officer based on observations during the fire and one or more subsequent site visits that may include interviews with 'eye witnesses', occupants, owner/manager of the site, neighbours, etc. In both the cases, the depth and thoroughness of investigation is guided by the extent of damage caused by the fire. Fires that result in 'nil' or 'minor' loss may end up with a one line report with very little investigating. Disastrous outcomes, on the other hand, may see the setting up of 'panels' or 'committees' comprising senior officials.

It is for governments and courts of a country to identify agencies as competent to investigate fire incidents. At the same time, it should be recognised that fire is a complex phenomenon and a fire incident should be investigated by people with good understanding of fire science, fire behaviour and fire behaviour of materials. They should also have deductive skills to reconstruct likely sequences of events using their knowledge of fire science. Additionally, in the case of industrial fire incidents, investigators must have knowledge of the activities carried on and the processes used at site and the hazards thereof. Interviews with witnesses (both eye witnesses and informed persons also known as expert witnesses) are an integral component of fire investigation and this requires both training and experience for interviews to bear results. To perform all these functions, fire investigators should have a range of traits and skills including:

- An open mind, focussed and determined but without preconceived notions about the incident.
- Keen observation together with an analytical mind.
- Strong base of fire dynamics.
- Good understanding of the fire behaviour of materials.
- Knowledge of activities at the location and associated fire hazards.
- Pleasant interviewing techniques, patient and with good listening skills.
- Photography and selection of appropriate subjects.
- Gathering, preserving and interpreting evidences.
- Deductive skills to re-construct the pre-fire scenario and sequence of events based on collected evidence.

Despite education and training, a single individual is unlikely to have all these skills and traits. Hence, particularly in case of major fire incidents, fire investigation is best done through a team of specialists; in simple, 'open and shut' type of cases where things are easy and straightforward, a single investigator alone may be sufficient. In any case, immediate pronouncement of a 'cause' of fire by 'experts' during or soon after a fire must be shunned.

In India, agencies called to investigate fires are diverse. The scope, purpose and depth of investigation also vary accordingly. In large, established industrial units, investigating teams may comprise in-house professionals from different functional areas. In major loss events, industry teams like those constituted by Oil Industry Safety Directorate (OISD) may be formed. Government agencies like the factory directorates of the states also conduct their own investigations and the police may step in particularly if there is loss of life. Where the public fire services have attended a fire scene, they investigate the incident. Insurers investigate fire incidents for the purpose of settling or refuting claims. Managements may also have incidents investigated by specialist third parties with the aim to improve their in-house systems. At the other end of the industry spectrum, fires in small scale units whether isolated or located in multi-occupancy industrial estates may be attended by public fire brigades and, in such cases, the fire services and insurers conduct investigations to meet their own requirements.

Often, investigators develop their own styles of investigation by observing their mentors. At other times, people in different professions are called upon to participate as members of fire investigation teams. This leaves much scope for misinterpretation of the origin of a fire incident or the likely sequence of events. Collecting and analysing evidence are important components of incident investigation. In addition, to catch the scene as closely as it was at the time of the incident, site evidence should be collected expeditiously. This scenario needs to be modified with formal learning programmes delivered by specialists from different related disciplines.

11.5 WHEN SHOULD AN INVESTIGATION BEGIN?

Investigation of a fire should start at the fire ground when details about flames, emergence of flames through openings, fire fighting operations including rescue and salvage can be observed. Photographs of the scenario taken by trained photographers can support any recorded observations. This can be effectively done if the attending fire crew has a trained and dedicated member. However, the investigation cannot be based on fire ground observations alone and subsequent visits to observe and rummage through debris are generally required. Talking to people present at the fire scene can help to get eyewitness inputs. The investigator should be cautious about conflicting information given by different individuals and (s)he should be prepared to dig further to establish the ground truth.

People are attracted to an outbreak of fire and, larger the fire, larger is the crowd of observers. People may also try to help fire fighters in their work. At times, the presence of crowds of people can pose obstruction in the operation of attending fire appliances and police help may be required for crowd control as well as to restrain people from getting close to the fire scene. The fire ground is also the place where people converse and comment. Fire crews need to observe people and pick up any helpful information. Investigation will necessarily involve talking to eyewitnesses and 'expert' witnesses and this should begin with minimum delay as people's perceptions and opinions can change over time. However, interviews with persons injured in a fire should be taken only when the victim has overcome the trauma.

While it is important for investigators to visit a fire site soon after an incident, it is to be remembered that fires can (and do) generate toxic products as we have discussed in an earlier chapter. In the case of enclosed spaces, the atmosphere inside may not be safe to enter for quite some time after the fire has been doused. Rummaging through debris may also release harmful vapours. For such instances, a rule of thumb is that one should allow twenty four hours after a fire has been doused to start site investigation. It is also important that investigators do appropriate respiratory protection during their visits to the site and particularly while entering enclosed spaces.

11.6 EVIDENCE

Evidence is what all investigations are based on. Evidence provides the building blocks that may be used to reconstruct a scenario and the possible sequence of events. Using all items of evidence, an investigator can prepare a report that can be understood and accepted by others. For instance, insurers decide the tenability of claims based on the reports of surveyors, their own engineers or even specialist third party investigators, the police may apprehend suspected arsonists. Collecting, analysing and interpreting evidence are vital activities for an investigator and the skills of an investigator relate primarily to gathering and interpreting evidence. In the case of major fires with extensive damage, much physical evidence is lost and finding out relevant and reliable evidence can be a major challenge for an investigator.

The foregoing discussion may have given the impression that 'evidence' implies physical features or items only. People have tried to segregate evidence into different groups. Irrespective of different groupings, evidence will comprise the following ***4 P'***s:

- Location and layout (***Position*** or scene).
- Equipment, materials, tools, etc. involved (***Parts*** or sequence).
- Witnesses — eye witnesses and other informed people (***People***).
- Documents (***Paper or soft data***).

Legal terminology divides evidence as ***direct*** and ***circumstantial***. 'Direct' refers to what is obvious while 'circumstantial' includes implications that may help either to establish a line of thought or to explain a thought process and can help establish a fact. Circumstantial evidence should be logical and reasonable and not based on speculation or conjecture. All types of evidence are important but it depends on the skills of investigator(s) to be able to focus on their interactions and establish linkages to develop an event sequence that possibly led to the final scenario. In this process, investigators must consciously try to ensure consistency.

Some examples of different types of evidence are given below.

11.6.1 Scene Evidence

Scene evidence can mean a variety of things including:

- Type of construction, occupancy and layout of the premises.
- Accessibility of the premises to fire crews or to likely miscreants.
- Observations made of flames and fire growth.
- Smoke and burn marks inside and outside the building.
- Open passages/interconnections inside the building.
- Broken window panes.

- Availability of fire protection appliances and their use.
- Presence of people in premises at the time of incident.
- Photographs and sketches of the above.

11.6.2 Material Evidence

Material evidence can cover a wide variety and may include:

- Materials of construction of the affected premises.
- Contents of the premises (raw materials, intermediates and finished goods).
- Equipment and activities carried out in the premises.
- Debris.
- Probable seat of fire.
- Possible source of ignition.
- Likely path and reasons for fire growth.
- Incendiary materials like accelerants.
- Photographs of the above.

11.6.3 People Evidence

People may be involved in different ways with the activity, location or incident management. They can include:

- Fire fighters and other emergency personnel who attended the call.
- Injured persons (if any).
- Other eye witnesses including employees where applicable.
- Management of the premises.
- Designers and vendors of the process/activities.
- Past employees
- Insurers
- Forensic scientists
- Expert witnesses (where the system exists)

11.6.4 Paper Evidence

Documents are always associated with an activity or site. In today's world, documents can be both 'paper' and 'electronic'. These can include:

- Standard operating procedures.
- Log sheets and log books.
- Parameter charts and printouts.
- Emergency management plans and training records.
- Inspection and maintenance records.
- Past incident histories.
- Financial records.
- Insurance policies and claim records.

An investigator must begin and continue through his consciously work without any bias. (S)he must concentrate on collecting evidence of different types, patiently analyse them and never jump to conclusions. The following sections discuss methods of collecting and analysing evidence.

11.7 WITNESSES

People referred to as witnesses may be eyewitnesses or others who have knowledge of the place, activity or systems in use there and who may be known as expert witnesses. An investigator must have an open mind and remain unbiased at all times including during interviews with witnesses. When collecting 'people' evidence, investigators should remember that

- Conduct interviews as soon as possible after an incident. Interviews of injured persons may need to be deferred.
- Conduct interviews individually and formally.
- Put the interviewee at ease.
- Obtaining people evidence should be more of listening and limited talking.
- ***Interview*** and do not ***interrogate*** witnesses unless arson is established through other forms of evidence.
- Allow the interviewee to give his/her statement. Ask questions only to clarify the contents and provide details of statements made.
- Never ask leading questions.
- Statements made by witnesses and clarifications obtained must be put in writing with date, time and place of interview recorded with signatures.

Immediate interviews help prevent witnesses being tutored. Persons involved in the activities of an affected facility at the time of the incident and attending emergency personnel are primary eyewitnesses. Fire fighters, patrolling policemen, neighbours and passers-by can all be eyewitnesses.

Expert witnesses help investigators understand details of an activity, the premises, materials used and produced, hazards involved, personnel, history, etc., at a location. People in the know of past experience such as insurers, loss surveyors and investigators of earlier incidents can give useful inputs. All such information can help an investigator logically reconstruct a scenario and the likely sequence of events in the incident.

11.8 BURN PATTERNS

A fire generally originates at one point where it leaves a footprint and fire investigation involves identifying this point. However, this may not be possible in some cases like that of fires in open surface liquid fuels spills. As a fire grows and convective currents are set up, burning progresses in the shape of the letter **V** often referred to as an inverted cone and this is evidenced on nearby vertical surfaces which, in the case of buildings, are doors, walls, furniture surfaces, etc. The rule of thumb is to try and locate the low-level burn marks and bottom of the **V** and hence the point of origin of the fire. However, there are exceptions to this as a curtain or drape might start burning at a higher point and then drop down to falsely indicate a lower point of origin than the real on. Similarly, materials like poly tetra fluoro ethylene (PTFE), brand name Teflon, can melt on heating and flow down slopes or stairs and once again the bottom of the **V** may be below the actual origin of fire. Such instances can mislead an investigator. Sometimes **V** patterns may appear on the outer side of windows and the lowest window where this is seen in the case of a multi-storeyed building should be noted.

In an earlier chapter, we discussed the growth of a compartment fire. The rising plume from a buoyant diffusion flame heats both materials that come in its convective path and other materials in the area by thermal radiation. This heating of walls and other surfaces by the rising plume causes discolouration on them. Some materials like wood surfaces, emulsion painted surfaces, plastic laminates or wallpaper may blacken, char, melt or deform. Investigators should use these signs to track back to the starting point of the fire as well as the path of fire spread.

Burn patterns may indicate the position (open or closed) of doors, windows or ventilators at the time of the fire. A clean section on an otherwise soot covered or burnt wall indicates that some objects which were present during the fire had been removed afterwards. If there are several points of origin that show burn patterns of equally intensity, it may indicate deliberate ignition. An investigator must photograph burn patterns that may support her (his) inferences.

11.9 SKETCHES, PHOTOGRAPHS AND DIAGRAMS

The saying goes that a picture can speak a thousand words and this is very true in fire investigation. A plan/layout drawing of the scene is important. In the case of industry, a process flow or work flow diagram could be useful. Sketches and photographs help to document physical evidence of the fire scene and support inferences drawn by an investigator. Appropriately captioned and numbered photographs can be significant evidence particularly if the scene needs to be cleared for reasons like resuming normal activity or to demolish unsafe structures. A word of caution: courts may not accept photographs taken with digital cameras as there is scope to manipulate and doctor the results. Importantly, though, an investigator must know what to capture on camera and how best to do it. Photographs should be arranged in a sequence that will support the sequence of events and provide some continuity for the reviewer. Using standard measures like a foot rule or metre rod when photographing a spot can help providing some idea of dimensions.

11.9.1 The Overall Picture

Photography of an affected site should begin with panoramic shots giving an overview of the location or facility relative to adjacent occupancies, approach roads, etc. A map of the area could supplement the photographs. More overall shots closer to the site can indicate overall fire damage, access routes and openings, fire hydrant network and risers. Close-up shots should be taken where it helps to draw inferences. Photographs of burn patterns on outer walls of buildings help to show paths of fire growth and could even point to the origin of the fire. Photographs of damage of window panes can indicate the status during the fire. The location where broken glass has fallen should be photographed as this can indicate whether it was broken from outside or from inside. Photographs taken from an elevation can show the layout of the premises and structural damage. This can be very helpful in the case of open industrial plants. These can be juxtaposed with pre-damage photographs to illustrate the extent of damage.

Videos taken by the media or others can be obtained to bear out information of flames, areas of intense burning, exit points of flames, etc.

11.9.2 The Detailed Picture

Next the investigator has to focus on detailed photography. Like a map for the overall picture, a layout plan of the premises can help to correlate with both photographs and text of the report. In the case of single family housing or a small store, such a plan may not be available and detailed sketches may be made. As mentioned earlier, photographs must be numbered, annotated and arranged sequentially. As one may land up taking more photographs than are used in the report, the ones that do not serve any purpose in interpretation should be excluded from the final report.

Signs of likely use of accelerants, areas of intense burning, prominent burn patterns, both sides of doors carrying burn patterns to capture damage to handle, latches, signs of break-in (if any), etc. should be photographed. Damaged electrical apparatus, particularly if burn patterns are noticed near them should be observed closely. If an electrical appliance or equipment appears to be the point of origin, it should be photographed from a distance and close to the point. The appliance itself is a piece of evidence and must be preserved for closer examination.

11.10 MATERIAL OR SEQUENCE EVIDENCE

Visits to a site ravaged by fire can be very disquieting. Yet it is from this debris and destruction that an investigator must find meaning in the form of material or sequence evidence. Materials of construction of the structures and equipment, materials used, processed, stored or produced, activities and parameters thereof can all be included in this class of evidence. Material evidence should help identify the material first ignited and further sequence indicating growth and path thereto.

Debris samples requiring detailed analysis must be collected in clean sample container and labelled appropriately. Cigarette stubs in an area supposed to be a 'no smoking' area is vital evidence. Items such as space heaters, burnt pieces of drapes, curtains, damaged electrical appliances and wires can also be items of evidence. The presence of some accelerants may be obvious due to their typical odour. A word of caution — one must be careful about trying to detect vapours by smelling debris as the vapours may be toxic.

Evidence of any accelerants used may be found by using samplers like Draeger tubes but even detection devices cannot differentiate between accelerant vapours and pyrolysis products of combustion. Advanced and expensive technology of gas liquid chromatography (GLC) can give precise results in debris analysis even if only trace amounts of accelerants are present. Trained sniffer dogs can also be used to detect the presence of explosives or accelerants.

Some of these items may need to be examined in detail in forensic laboratories. Material like piles of paper, heavy timber, carpet and concrete can retain accelerants for long while thinner combustibles like sheets of paper and wood veneer burn with any accelerants they may have contained.

The inferences drawn from material evidence should contribute in the overall exercise of locating the probable seat of fire and possible source of ignition. The American Society for Testing and Materials (ASTM) has published a set of standards on sampling and analysis from fire debris.

11.11 RECORDS OR PAPER EVIDENCE

All occupancies will have at least some hard physical records although much today is kept in 'soft' form. Even homes have records of ownership, bank accounts, insurance policies. Educational institutions will have records of student numbers. Business will have volumes of records including log sheets, inspection reports, financial performance reports, loss incident history, etc. All such records, whether kept on site or elsewhere, can provide vital clues and information about the physical and financial status of an occupancy. These can be part of circumstantial evidence. Burnt paper, on the other hand, can be direct evidence. Individual sheets of paper can burn out totally, but tight stacks of paper may burn at the periphery and remain largely unaffected inside. Burnt papers can point to the path of fire growth. Papers also absorb any accelerants used and this can show up in forensic analysis. The fire plume established in a fire can throw up pieces of burning paper and spread the fire.

11.12 INTERPRETING EVIDENCE AND REPORT WRITING

After study of the history of the premises and similar locations, site visits, witness interviews, etc., investigators will have a lot of evidence and they have to sieve out what is useful from what is not. This is a critical activity and investigators have to use their skills to interpret evidence and reconstruct the scenario. Support of specialists like forensic scientists may be required at this stage.

The efforts of an investigator should culminate in a report that includes the following:

- Activities at the location including materials and processes used, hazards involved, manpower, management and past fire experience.
- Abnormalities (if any) at the time of fire.
- Details of detection, emergency response including constraints, attending manpower and appliances.
- Sequence of events as reconstructed starting from material first ignited, source of ignition, the path of growth, etc.
- List of photographs, sketches, etc. used in the report.
- Witness statements.
- Estimates of loss and damage.

11.13 'ELECTRICAL' FIRES

When electrical energy is converted to other useful forms of energy, heat is produced. Sometimes heat is the desired output as in ovens or electric furnaces. A glowing incandescent bulb gives light and generates heat; an electric motor provides mechanical energy and generates heat. The transport of electricity through conductors also generates heat. A fault in any part of the system can increase the current flowing and the heat generated till the appliance, the connecting wiring or its insulation or the receptacle may either be damaged or burn out. A circuit breaking device such as a fuse of appropriate rating should prevent this from happening.

The frequent reference to 'electrical' causes in fire statistics may not be quite appropriate but fires often do involve electrical systems and appliances. This may be due to defective equipment, misuse, poorly maintained appliances, appliances left unattended in the 'on' condition, circuit breakers of higher rating, etc. Clothes, paper or drapes getting ignited by an electric heater or a hot electric lamp are other similar examples. It is certainly wrong to refer to any of these as fires of electrical origin.

An ***electrical fire*** can rightly be described as a fire where an electric current or an electrical fault is the likely source of ignition. Some examples are:

- Very high voltages and currents due to lightning strikes generate so much heat that nearby combustibles can get ignited.
- The current flowing through a conductor finds greater resistance at loose connections and this generates excess heat, which can start a fire at a terminal like a junction box or inside a switchboard. Such fires have occurred even in cars where the voltage is just 12 V.
- Intermittent arcing (spattering) or 'series' arc occurs when conductors are not in direct contact but an arc current along the carbonised insulating material between conductors as at a receptacle. Being intermittent, circuit breakers may not operate despite high current flow. This may continue for some time and result in fire.

Heat from incandescent lamps can ignite adjacent combustibles. Ballasts in fluorescent lights can overheat due to internal faults and start fires. Any material that has dried out can overheat and ignite in a microwave oven. Metal objects inside a microwave oven can cause severe sparking and lead to fire. Combustible material coming close to a heat emitting component of an appliance can start burning. Air conditioners draw high currents for extended durations and can cause fire by a hot connection or an overloaded circuit. The use of combustible thermal insulation increases chance of fire.

Electrical fires most commonly occur at 'weak points' in the system such as where cables have been cut and joined, or where they have been cut, and connected to a power point, light or similar usage. Moisture or dirt can build up across 'live' terminals, or, the connections may become loose with time and cause localised heating.

When determining if an electrical appliance was a likely source of ignition, it should first be checked if the appliance was connected to the supply at the time of the fire.

Another source of fire due to electricity is from ***electrostatic discharges*** that can be generated in a range of situations including lightning strikes. ***Static electricity*** is generated whenever two dissimilar materials repeatedly make and break contact. This happens in the passage of materials like passage of finely divided solids or in the movement of liquids through pipes. If sufficient potential difference is built up and a spark of adequate energy is generated in an atmosphere of combustible concentration, a fire (or even explosion) may result. The following five conditions must occur simultaneously for static to be a source of ignition:

- A flammable atmosphere must exist.

- There must be static generation.
- There must be static accumulation.
- The static accumulation has to discharge.
- The discharge must have sufficient energy.

Before naming static electricity as the source of ignition in any instance of fire, an investigator should confirm that the above five conditions did occur simultaneously.

11.14 CONCLUSION

Fire investigation has been likened to archaeology but, unlike in the latter, a fire investigator cannot choose the site. Both, however, aim to interpret history and the likely course of events that led to the present scenario. The skills required in fire investigation are many and a team rather than a single investigator may be required to try and reconstruct the scene. It requires much dedication too besides an open mind. It might be tempting to ascribe a 'cause' to such items as 'electrical short circuit' to cut short the investigation process. Investigators must consciously try to avoid such inferences and arrive at actually potential sources of ignition and path towards the final outcome. Finally, the advent of computer fire models has impacted the field of fire investigation too. A brief description thereof is given in Chapter 12, Modelling for Fire Safety.

QUESTIONS FOR REVISION

1. What is the significance of a V burn pattern in fire investigation? Describe some scenarios where this can lead to misunderstanding the point of fire origin.
2. What are the 4 P's of evidence in the investigation process? Give examples for each of them.
3. How can burn patterns, smoke marks, charred timber, etc. indicate path of fire growth?
4. What is an 'accelerant'? Describe ways in which the use of accelerants at a fire scene can be evidenced.
5. What are the conditions that must simultaneously exist for ignition by static electricity?
6. Contrast the terms 'interview' and 'interrogation' vis-a-vis investigation. What are the suggested ways in which witness interviews should be conducted?

12
Modelling for Fire Safety

We know from chaos theory that even if you had a perfect model of the world, you'd need infinite precision in order to predict future events.

—Nassim Nicholas Taleb

The idea that a computer model might partly replace experimentation is both dangerous and attractive.

—SFPE Handbook of Fire Protection Engineering

12.1 INTRODUCTION TO MODELS

Models are said to describe our beliefs about how the world functions. They are not reality but should be good enough to explain or understand reality. Models may be descriptive, physical or symbolic. Descriptive models may be explained as metaphoric that use the language of comparison.

Physical models of larger items of use have always been used as children's toys, be they aeroplanes, automobiles, human adults, animals or even objects of warfare. All these models are "scale down" models that aim to resemble reality and give children a 'feel' of the adult world. Physical models are also designed and created in laboratories with a view to understand different phenomenon and to scale up to sizes of practical utility.

Flight simulators and industrial process simulators are other big examples of models that are used extensively as very effective training aids. Simulation is especially useful in scenarios where human intervention is crucial as it offers opportunity for improving model-operator interface utilising virtual reality techniques to give real-like experience without exposure to hazards and at low cost. Another type of physical models are analogues where comparisons are used to explain factors or phenomena such as the level of water in a reservoir and rate of outflow are compared with voltage and current, respectively, in electricity. Physical models may also be represented through laboratory set-ups wherein experimental fires are created and their effects as well as their developments are observed and inferred.

From another perspective, people often try to draw conclusions from prior life experiences and predict the future course of events based on such "conclusions". Such thought, associated analysis and decision making processes can be formalised to be interpreted as a type of model. Further, when models of this type can be interpreted and justified with the help of mathematics or mathematical equations, we have what is known as *mathematical models*. As a precise and concise language, mathematics has defined rules for manipulations and, in today's world computers can be used to perform most complex numerical and iterative calculations. A mathematical model may be described as an abstract method that uses mathematical language to describe the behaviour of a system.

Probabilistic models provide quantitative estimates of the probability of occurrence and will be kept outside the purview of our discussions. In this chapter, we will confine our discussions to deterministic mathematical models which necessarily means the type that helps us analyse outcomes so as to understand their underlying causes, recognise paths thereof and be able to forecast potential path(s) forward. Significantly, we will also exclude any reference to fires in the open of which much research and modelling work has been done with respect to forest fires in different countries including India.

12.2 MATHEMATICAL MODELS

Mathematical models may be described as groups of mathematical equations derived from basic physical principles, real-life experience, or experimental data. They have been used in many disciplines of science and engineering as also in the social sciences like economics, dispersion of released substances, population studies sociology, political science and fire dynamics. They may take many forms such as statistical models, differential equations, dynamical systems etc. However, it should be recognised that they involve substantial elements of assumptions and compromise which include

- Determining the significance of various aspects of a system and deciding to include or exclude them in a model
- The complexity of equations chosen, the quantum of mathematical manipulations considered worthwhile as well as the precision desired in the output
- The time and computer resources available for the exercise

Mathematical models can take many forms such as dynamic (that move with time) systems, statistical models, differential equations and game theory models. The issues to be addressed through a model are often classified into either *black box* (when there is no prior information available) or *white box* (also called glass box or clear box) model where all relevant and necessary input information is available beforehand. These, however, remain only as nomenclature as normally most systems lie between the two model types.

Laboratory and pilot-scale models have been used extensively to understand the fire process and its dynamics and then develop mathematical correlations thereof. Once sufficient confidence is obtained about these correlations by validating them against different experiments as well as against actual occurrences, there is scope to predict fire events, sequences and scenarios to a great extent such that appropriate detection, fire control and life safety systems can be designed, selected and implemented. The correlations can also be used to explain incidents and re-construct scenarios in the process of investigating fire incidents.

12.2.1 Stages of Modelling

Models go through four stages of activity viz.,

- Build
- Study
- Test and
- Use

(This closely resembles *Deming's PDCA*, i.e., Plan, Do, Check, Act cycle.) If any changes are made to a model, the study and test stages need to be repeated. The process of repeated iteration is typical of mathematical modelling and this is very useful for improving understanding about the functioning of a system.

12.3 THE ADVENT OF MATHEMATICAL FIRE MODELS

Howard W Emmons (1912–1998), who served as a Professor of Mechanical Engineering at Harvard University for decades, was (among other things) known as the father of modern home fire research. The Harvard "Home Fire Project" led by Professor Emmons beginning in the mid-1960s systematically documented the scientific basis of the Harvard Computer Fire Code, a model predicting the development of fires in enclosures. Known as *Fire Code Version1.0*, this was the prototype for many later fire models. In a tribute published on the occasion of his death in November 1998, the Harvard Gazette said that his laboratory experiments "pushed the prediction of fire behaviour out of the realm of badly educated guesswork and into the world of precise mathematical modelling.... His efforts bore fruit in models such as the constantly evolving Harvard Computer Fire Code (which predicts how fires spread in buildings)...". Emmons was instrumental in establishing a fire research laboratory at Factory Mutual Research Corporation (now FM Global) where full-scale tests could be conducted to validate the theory.

Fire models can be described as engineering tools ranging from simple one-line correlations of data to highly complex models primarily intended to answer questions about the effects of a fire scenario. We are all familiar with examples of simple algebraic equations that involve only a couple of unknowns and which can be resolved using fundamental knowledge of algebraic operations. Similar is the case of simple processes involving only a limited number of parameters connected in relatively simple ways as was the case when the complexities of fire dynamics were understood less than they are today. Till such time, experimental studies were carried out under controlled conditions to learn about the fire phenomena and their various effects. Mathematical equations to represent the monitored parameters were developed by researchers in leading test laboratories and the equations developed were generally seen to be true.

Fire modelling involves explaining how fires start and spread, causing changes in the environment (closed such as in single or multiple compartments as well as open such as woodlands or built environments) caused by a fire incident that varies in intensity with time through a combination of physics-based simulations and mathematical equations. Advanced fire modelling complements large-scale testing by helping develop better tests and thus learn more from the results.

However, with the growing number of parameters recognised as significant players in fire dynamics and the increasing complexity of structural geometry or open spaces where fires have occurred, it became increasingly difficult to design experiments or derive equations that could be representative of real fire scenarios. There was also the factor of the cost of experiments which often were required to be repeated for confirming outcomes. Fortunately, the advent of computers and the facility of complex computing programmes helped to replicate actual fire development at different locations. This also facilitated solving complex algebraic equations that required iterative manipulations to arrive at what may be termed as "predictions". These "predictions" were validated against real events and the sequence thereof as and when they occurred. Sometime around the 1970s–1980s there was a growing interest in computer fire modelling both in the study of fire dynamics and in building for fire safety. The focus and factors of different models that were developed varied and selecting a model to use necessarily involve tradeoffs in modelling capabilities against requirements. The wide range of fire model types is summarised in Figure 12.1.

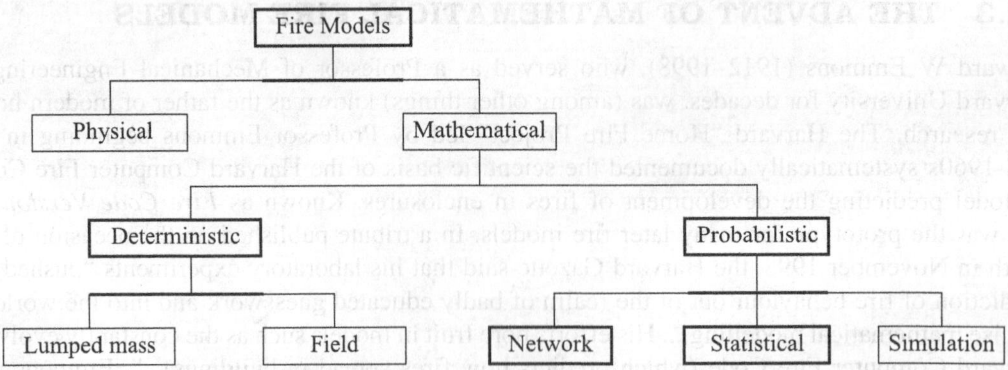

Figure 12.1 Range of fire models.

Various factors need to be considered before selecting a particular model depending on the purpose or use of the model. For instance, while designing a system, factors to be included are computational resources, time limitations, required level of accuracy, and most importantly, whether or not the governing equations and assumptions in the model are appropriate for the problem of interest. However, increased sophistication does not always translate into better results, as in many cases a simple model is adequate, or data required to use a more sophisticated model may not be available. In another scenario, for a given application, there may be one, multiple, or no fire models available.

12.4 TYPES OF FIRE MODELS

There are many models that have been proposed to predict various fire phenomena ranging from simple algebraic correlations to zone (or lumped parameter) models that divide a closed space into a small number of parts, to computational fluid dynamics (CFD or field models), which break up a space as a large number of discrete volumes. The three types of models from algebraic to zone and CFD are also indicative of increasing sophistication.

Heat transfer models are useful to analyse rise of temperatures in the surroundings due to a fire. This can help identify appropriate locations for sprinklers or other temperature-based protection devices or to assess fire-resistant character of the structure and its elements. Some other models focus on the movement and dispersion of smoke and other harmful products from a fire to determine safe egress of occupants. The heat and smoke profiles obtained as the output of these computations can be demonstrated in pictures or video animations of the fire condition and this can explain aspects of the fire scenario even to non-specialists.

Fire models can be classified in different ways depending on either their level of sophistication or on their projected end use:

- Fire models
  - — Algebraic models
  - — Zone models
  - — Field models
- Egress models

12.4.1 Algebraic Models

A simple analysis of fire dynamics has traditionally used algebraic models based on equations meant to assess fire characteristics such as flame height, heat release rate, plume and ceiling jet velocities, gas layer temperatures and depth, radiation etc. These equations have been developed by fire scientists and researchers either by using fire test data or by mathematically using physical and chemical properties and principles of fire dynamics with plausible extrapolations. They are proven models and useful to arrive under steady-state conditions, providing information at a moment in time. However, some algebraic models can also be used to provide transient data by solving the equation(s) iteratively over a number of time steps.

While algebraic models can provide answers to many questions of fire dynamics and basic fire phenomena they also have their own limitations. These include aspects such as that they are often tailored for specific simplified scenarios and their ability to predict developments may be inadequate when the actual conditions are different. Similarly, algebraic models generally provide information about a single property of interest, rather than a more detailed representation of a fire scenario. The ability of such models, therefore, to predict the various parameters of a fire scenario depends largely on the extent of match between the actual scenario and the original test conditions based on which the model was developed.

Despite these limitations, however, algebraic models are often used for their simple calculations to check the output obtained from more complicated forms of modelling. And even if the inferences from algebraic models may not precisely match those from more sophisticated zone or computational fluid dynamics models, they are often of the same order of magnitude. In case even the order is very different, it is advisable to analyse the reasons thereof and determine the model is indeed representative of the scenario or identify errors in the analysis that should be corrected.

When an algebraic model and its equations are found inadequate in terms of desired detail or accuracy of the fire parameters, the next step could be to use a zone (or lumped parameter) model.

12.4.2 Zone (or Lumped Parameter) Models

The concept of zone models is based on the assumption that when a fire occurs in an enclosed space, the characteristics within the space developed such that it could be broken up into a number of sub-spaces or "zones" each having somewhat uniform properties and over which algebraic equations could be used to determine different flow properties. Generally, zone models of a compartment fire divide the space into two zones (the first two in the list below) but up to four or five zones have also been proposed viz.,

- the hot upper layer that contains hot combustion products
- the cold lower layer containing ambient air and entrained air and sometimes
- a zone of the plume or ceiling jet
- the compartment walls
- objects in the compartment that may also be fuels.

In a two-zone fire model, the basic assumption is that hot gases will rise towards the ceiling of a compartment to form a hot layer in the upper regions and a cooler layer will be present in the lower regions of the compartment. The interface between the hot and cold zones keeps descending as the volume of hot gases increase with time. The upper zone receives both mass and energy from the fire and loses energy by convection or mass movement of gases through openings, by radiation to the floor, and to the surfaces in contact with the upper zone by conduction and radiation.

Although actually there are variations in conditions within each zone, the variations are relatively very small as compared to differences between the layers and, therefore permit use of this basic assumption while using zone models. There are significant mass and energy transfers between the zones. It may be mentioned that zone models are often used to monitor conditions during the transport of smoke and hot gases from the compartment of fire origin to other portions of buildings via connected volumes like emergency escape routes and in which case the intra-zone variations may not be very significant. The outputs of zone models depend on a user-defined characterisation of the design fire (Figure 12.2).

With respect to zone models, the possibility of a single zone scenario is often either overlooked or not contextualised. However, the scenario does exist in specific cases such as in post-flashover compartment fire. Similarly, single zone approximation has also been recommended when

- the temperature of the hot layer exceeds 500°C or
- the height of the hot layer extends down to 80% of the compartment height from the ceiling

(*Note:* Experimentally it has been seen that flashover occurs when the hot layer temperature is 500–600°C.)

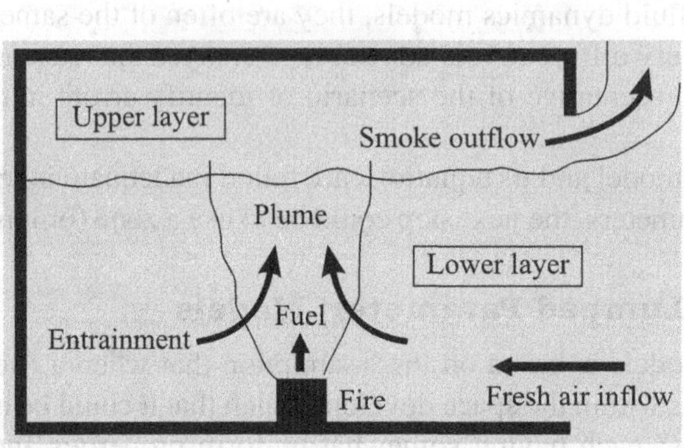

(a) Two zones of a compartment fire

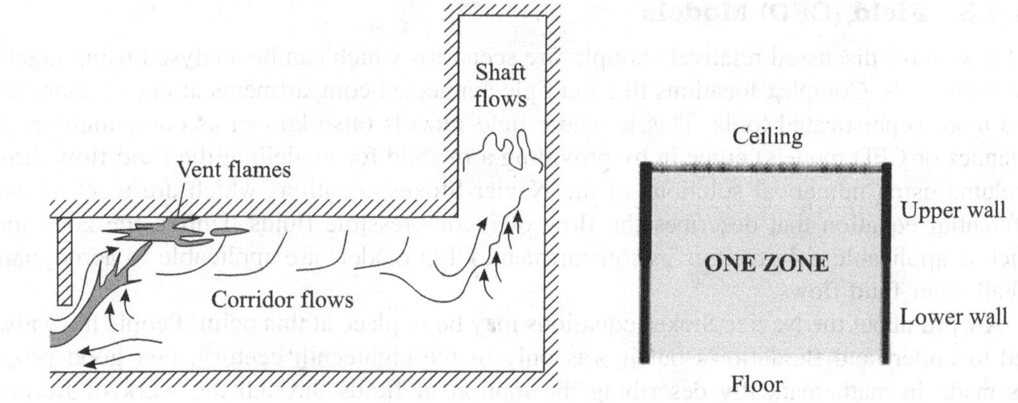

(b) Scenario if (a) is connected to a corridor and rising shaft (c) Single zone scenario

Figure 12.2 Zone concept in a compartment fire.

Advances in personal computer capabilities have made it possible to run some zone models even for multi-compartment fire scenarios. This in turn has made it possible to obtain more information than might be available than by using algebraic models. It also facilitates including a wide range of parameters that may affect the behaviour of a fire. Zone models, however, have their weaknesses too. For instance, the assumption of uniform zones is not very correct. It is also difficult (if not impossible) to model spaces of complex geometries trying to simulate them as equivalent rectangular volumes. Further, some basic zone model assumptions like the ascent of hot gases to the ceiling may not be true in the case of very tall ceiling height like atria where stratification can occur at different levels.

Some zone models have additional sub-models to simulate specific phenomena not inherently consistent with the zone assumption. For example, some zone models simulate detector or sprinkler actuation through the use of ceiling jet calculations. They may calculate the velocity and temperature within a fire room, but assume a different velocity outside the fire room as the result of a door plume. In the case of a space provided with sprinkler protection, the two-zone assumption breaks down when the sprinkler actuates. While this may be adequately covered through suppression correlations when a single sprinkler operates, such an approach may not be adequate for multiple sprinklers simultaneously operating.

Model inputs typically include compartment geometries, building materials including lining and furnishings, details of vents and other openings, contents and heat release rate from a potential fire. Some models also provide for openings to the outside, connected compartments etc. for heat losses to the walls and ceiling. Prevailing ambient conditions may also be included. Outputs may include activation time of fire protection devices and systems, time to flashover, physical characteristics of the upper and lower layers, and concentrations of combustion products.

Before closing our discussion in this section, there are two important points about zone models viz.,

- They are appropriate for compartment fire scenarios only and
- A primary assumption is that the fire is located at or close to the floor.

12.4.3 Field (CFD) Models

So far we have discussed relatively simple fire scenarios which can be analysed using algebraic or zone models. Complex locations like multiple connected compartments at one or more levels need more sophisticated tools. This is where field models (also known as computational fluid dynamics or CFD models) come in by providing a method for modelling the fluid flow through a volume using numerical solutions of the Navier-Stokes equations which are a set of partial differential equation that describes the flow of incompressible fluids. Unlike the zone model which is applicable only to fires in compartment, CFD models are applicable to aerodynamics and all other fluid flows.

A word about the Navier-Stokes equations may be in place at this point. People have always tried to understand fluid flows but it was only in the eighteenth century, that good progress was made in mathematically describing the motion of fluids through the work of Bernoulli, Euler, Reynolds, Poisson, Lagrange and others. Euler's equations describe the conservation of momentum for fluid of negligible viscosity and conservation of mass. The nineteenth century French engineer and physicist Claude Navier and the Irish mathematician George Stokes added the factor of viscosity into the Euler equations, which resulted in what is now called the Navier-Stokes equations. These partial differential equations describe the conservation of mass, momentum, pressure, species and turbulence and they form the basis of computational fluid dynamics. Work in the twentieth century brought further refinement in understanding fluid flows with theories of boundary layers and turbulence which are also fundamental elements of modern CFD.

Field models divide a compartment into a large number of small three-dimensional boxes called *cells* of dimensions ranging from centimetres to meters. These models are based on the basic principles of conservation of energy, mass and momentum. The computer programme facilitates the computation of heat and smoke movement between the cells over time. It is possible to find the temperature, velocity and gas concentrations within each of the cells at any point in time, also which are assumed to be constant within each cell (similar to the case of zone uniformity in zone models). Due to the larger number of cells, however, the conditions in the enclosure can be predicted in much greater detail.

Around the late 1970s and the 1980s, scientists and engineers began using computer models to calculate the effects of fire but the models of that time were constrained by computer resources available at that time. Widespread developments in information technology in the 1990s assisted complex calculations of fire effects and also the ability to graphically illustrate scenarios and inferences. Today's computer fire models utilise *computational fluid dynamics* (CFD) analysis to provide detailed calculations of the movement, concentrations, and temperature of smoke in a building. (Such models have been used in aviation design.) Heat transfer models have also been developed to analyse temperature profiles in a fire scenario. These are helpful to analyse structural fire resistance and to identify heat detector or sprinkler locations. When computer fire models with 3-D capability are used, conditions prevailing in different portions of a building can be visualised through pictures or video animations.

Mathematical fire models involve the principles of conservation of mass, energy and momentum across zones. In the case of field models wherein a large number (hundreds of thousands) of control volumes are addressed, the issue is addressed with the help of partial differential equations and the laws of mass, momentum and energy conservation are applied in each cell and balanced with all adjacent cells. The CFD models also have adequate flexibility to simulate a wide range of boundary conditions and the ability to model complex geometries. Additionally, most CFD software packages permit these calculations to be performed over a period of time to provide transient solutions. Thus, as compared to zone models, field models facilitate a finer level of resolution but this requires greater computational resources, numerous runs and extended simulation time which may be days, weeks or even months.

The temporal capabilities of field models permit addressing fluid characteristics like turbulence and buoyancy and further facilitate simulation of analysis of phenomena such as the spread of hot gases, smoke, thermal radiation etc. to simulate a fire-driven flow. They are capable of modelling pre-flashover and localised fires in complex geometries with smoke movement in multi-compartments.

12.5 FIRE MODELLING—APPLICATIONS

Computer fire models have found broad applications as their capabilities are getting wider recognition both before and after fire instance some of which are listed below.

- To study
  - — Fire growth and behaviour
  - — Time-line of fire outcomes
  - — Flame heights and size
  - — Temperature profiles
  - — Effects, heat release rates and heat fluxes
  - — Movement and flow of smoke and other fire products
- For fire prevention and protection
  - — Determine factors of fire resistance for life and building safety
  - — Evaluate performance or objective based design alternatives
  - — Establish adequacy or otherwise of prescriptive life safety codes
  - — Selecting and locating fire protection devices (detectors, alarms, sprinklers etc.)
- Egress modelling
  - — Study response of occupants to a fire
  - — Evaluate movement of people and occupant egress
  - — Determine effectiveness of smoke control and other protection systems
- Investigate and reconstruct fires scenarios (Remember fire destroys evidence necessary for modelling.)
  - — Analyse fire incident to identify stages and time-line
  - — Reconstruct possible scenarios from end-result
  - — Validate models
- During fire
  - — Predict possible growth of fire (Seldom done but which may be beneficial for fire fighters)
  - — Assess efficacy of installed fire protection systems.

12.5.1 Fire Reconstructions

A major use of fire models today is in fire investigation as an analytical tool, to study actual fire scenarios and to validate different models. Although much of the physical evidence is lost in a fire, information in the form of eyewitness accounts, forensic evidence, fire department reports etc. are still available to make the use of computer fire modelling for fire investigations. In the case of investigation, therefore, a computer fire model is more useful to supplement other information in demonstrating that a particular hypothesis is plausible or not and not every case will warrant the use of models. This is like investigators not knowing to use sophisticated analytical instruments but knowing the correct methods of collecting and preserving samples for sending to the laboratory for fire debris analysis.

In the case of fire investigation, computer fire modelling can be used for:

- testing hypotheses
- understanding fire behaviour
- safety of occupants
- timeline analysis, and
- validating or refuting post-fire damage.

It is not appropriate for someone to utilise a computer fire model to prove causation although the observed effects of an actual fire can be used to validate the hypothesis proposed in a model. Computer fire models will never replace a good on-scene investigation. Investigators need to be very concerned with those that believe that they can perform an investigation by sitting at a computer. It is important for all investigators to remember that computer fire models should be used as a tool to supplement an investigator's on-scene investigation and analysis.

12.5.2 Input Data for Use in Scenario Reconstruction

Computer fire models are increasingly being used in fire investigation and analysis although they do not preclude the use of on-scene investigation which also will provide some of the data required for modelling. The general input data needed for computer fire modelling that must be obtained at the scene are:

- ***Fireground geometries***—A primary and vital input required to recreate a fire scenario for use in computer fire models is to start with an accurate 3-D scene diagram that includes all geometric dimensions and details of contents with their locations. Details of all vents including HVAC and any other forced ventilation provisions including their capacities should be known.
- ***Lining materials***—The materials used to line walls and the ceiling as well as flooring or floor cover impact transfer of energy out of the compartment. Hence, such data are also important inputs for modelling fire scenarios. It is advisable to collect and preserve samples of these materials.
- ***Fuels and fire growth***—The types of combustible materials, their state, size, orientation etc. and their fire properties are vital inputs into a computer fire model as these will determine the heat release rate over time. Here too, it is advisable to collect and preserve samples of the fuels for further analysis.

- ***Fire protection elements*** – The types of all fire protection provisions such as detectors, sprinklers and other fire control systems elements as well as their locations must be identified on the 3-D diagram as referred in the paragraph on Fireground geometry above.

12.5.3 Egress Modelling

Needless to say, the most preferred ways to reduce fire-related fatalities should be preventing fires, reducing the number of fires and controlling their spread. However, if and when a fire does occur, evacuation is a primary life safety consideration. Providing (or absence of) adequate means of egress can affect two major direct costs: installed means of fire protection and fatalities or injuries. US statistics estimate that roughly a quarter of the over 2500 deaths in structural fires every year happen during evacuation.

As the name indicates, the primary purpose of *egress models* is to assess and analyse safe egress of occupants from a structure on fire. It is no wonder, therefore, that egress models have emerged as a good tool to plan and design for safe evacuation. Computer egress models are complementary to computer fire models in that they help estimate egress times in a building configuration based on various fire scenarios and prevailing conditions such as an exit being blocked or unavailable. Results from a fire model form the basis of egress models.

Egress models can range from simple to very complex with the latter capable of analysing the movement of thousands of occupants in complex buildings and structures and egress through the building's floors and paths. Information provided by egress models can be used to create 3-D video animations depicting egress from a building or structure. Depending on the complexity of the building in case, algebraic hand calculations to computational evacuation models are used to assess safe egress requirements. A basic assumption in such analyses is that occupants are standing at the doorway of the egress path on each floor when evacuation begins. Such analyses focus mainly on constraints along the path to safety and arrive at the time for the occupants to reach outside.

However, a very difficult area to negotiate concerns the behaviour of people which is difficult to predict with any degree of precision. To an extent, these are guided by prevailing conditions of heat and smoke, training and experience, fear perceptions, pushing and overtaking etc. Hence efforts are on to include stochastic inputs in egress models. Another basic facet to be included in egress models is the mobility of occupants such as in the case of hospitals, kindergarten schools etc.

An important area where egress models can be helpful is where estimated safe egress requirements are in variance with norms laid down in prescriptive building and life-safety codes may either not be adequate or even relevant under actual fire scenarios. For instance, codes often limit permitted travel distance from any point in a building to an exit apparently to represent time taken to get out from a fire-affected building. On the other hand, computer egress models can provide justification for longer travel distances by improving design features like wider escape paths or improved design.

As a point of information, egress models have been developed specifically for ships, aircraft, tunnels.

Many different egress models have been proposed with a range of parameters representing various scenarios. These include differently-abled people, the sick, old and infirm, children and people asleep or awake. While these physical differences may be accounted for in models, human psychology remains an enigma and ways in which people will respond and react to fire and its effects cannot be firmly stated. People go through several different phases from receiving a warning, perceiving a threat, personalising the risk, and deciding to act to protect self and others and property in response to a disaster. Much also depends on socio-cultural exposure and experience, disaster training received etc.

Some egress models that have been developed and have found use are EXITT, Evacnet, SIMULEX, EXODUS etc.

12.5.4 How Can Computer Fire Modelling Help Fight Fires?

Wind is a major factor that contributes to fire spread in the open and more so in the case of forest fires. Places that experience forest fires regularly have fire fighters trained to address the impact of wind in their tactics. However, this aspect of training may be missing in structural fire fighting even though wind can and does impact fires in the built environment. The National Institute of Standards and Technology (NIST) has used detailed computer fire models to determine the critical impact of wind in relation to the behaviour of fire and safety of firefighters, even indoors. The model with wind indicated that the fire followed a wind-driven flow path that rapidly increased temperatures (over 260°C) in some areas where firefighters were working and which was absent in the model without external wind. Thus, the temperatures and conditions in which the firefighters were working, were much less dangerous in the scenario without wind.

12.6 REPORTING AFTER A MODELLING EXERCISE

Like any other exercises, a report is expected at the conclusion of a modelling exercise. The report should include the following:

- Name and version of the model used
- Sub-models used for analysing different aspects
- Assumptions made about the design fire, compartment features, ambient conditions etc.
- Summary of results

12.7 VALIDATION OF COMPUTER MODELS

The acceptance of computer models is largely subject to confirmation by their validation but validation per se is not a "fit and forget" task but is an on-going process in the life cycle of the model. The process involves continuous monitoring of four types viz.,

- Testing components of the model and
- Three stages of validation that include
  - — Functional validation
  - — Qualitative validation and
  - — Quantitative

It is the last stage of validation that could be challenging with inadequate experimental data.

12.8 SOME POPULAR MODELS

Several research institutions and universities have done extensive work relating to both developing computer fire models and in their various applications. Prominent among them are NIST (National Institute of Standards and Technology) through its Building and Fire Research Lab, United States, FM Global, BRE Global (formerly the Building Research Establishment and the Centre for Fire Safety of UK), BRANZ (formerly Building Research Association of New Zealand), Harvard University, University of Edinburgh, Greenwich University (UK), Lund University (Sweden), Worcester Polytechnic Institute (USA) and University of Canterbury (New Zealand).

Many of the different models that were developed as the dynamics of fire and its growth were increasingly recognised and also as computational capabilities grew have been archived and are presently only of historical importance. Some of these are

- HARVARD V
- ASET (Available Safe Egress Time) and ASET-B
- FIRST (FIRe Simulation Technique)

12.8.1 NIST Models

The Fire Research Division of the National Institute of Standards and Technology (NIST) of the US has been a leader in developing computer-based fire models. Two widely used models developed by NIST are:

- Consolidated Fire and Smoke Transport (CFAST) and
- Fire Dynamic Simulator (FDS).

Both the models have been upgraded continuously to incorporate new learning of fire dynamics and the strides in computer and information technology.

The *Consolidated Model of Fire Growth and Smoke Transport (CFAST)* is a "two-zone" model developed in the 1980s with characteristics described above. It has the advantage of producing a model in just seconds, but the disadvantage of providing only average compartment temperatures.

Typically CFAST predicts temperatures and concentrations of fire products in the two (hot upper and cooler lower) layers and interface heights in a multi-compartment structure. The model is widely used by fire investigators, fire safety engineers, architects and builders to simulate the impact of fires (actual or potential) and smoke in a particular building environment.

The CFAST package includes the *Smokeview* software which presents coloured, 3-D animations of the simulation of a fire scenario in terms of heat release rates, temperatures, flows, various gas concentrations and growth and movement of smoke. The software also indicates, actuation of protection devices across multi-room structures.

The *Fire Dynamics Simulator (FDS)* is a computational fluid dynamics (CFD) model of fire-driven fluid flow developed at NIST that was first released in the year 2000 and has subsequently been upgraded in several versions. It provides much greater detail of the fire and the gas flows than CFAST, but the runs can take hours or days to perform.

FDS requires the subject building to be divided into 3-D computation cells and the software solves numerically an appropriate form of the Navier-Stokes equations to numerically compute the density, velocity, temperature, pressure and species concentration of the gas in each cell. The model tracks the generation and movement of fire gases based on the laws of conservation of mass, momentum, species and energy.

Inputs required by FDS include:

- Structure geometry and details of contents and furnishings
- Computational cell size
- Location and heat release rate of the fire

The right selection of input parameters greatly impacts a simulation but, in many cases, all details of contents and furnishings at a fire scene may not be known. This is particularly true in a post-fire situation. Keeping this constraint in mind, a range of values is typically used as input into the fire model. This necessitates that numerous runs are made to determine the parameters that best align with the physical effects, images of the scenario, fire timeline, witness statements etc.

Smokeview (SMV) is a visualisation program used in conjunction with both CFAST and FDS to graphically represent the simulation results.

FDS+Evac is the evacuation version of FDS wherein evacuation simulations can be fully coupled with the fire simulations.

12.8.2 BRE Global Fire Models

CRISP is a two-layer zone model for smoke flow in multiple rooms which is coupled with a detailed model of occupant behaviour and movement. It is thus a building fire risk assessment and egress model. The sub-models in it represent physical 'objects' (rooms, vents and openings, protection devices, contents, etc.), hot smoke layers and occupants.

CRISP is a BRE model that can be used to simulate aspects of a fire scenario including fire growth, smoke movement, detection and suppression, human response and evacuation and can thus be effectively used as a fire risk assessment tool. The model includes algorithms for human behaviour and hence can be used separately as an egress model. The Monte Carlo controller handles all the physical objects as well as all inputs and outputs and calculations are carried out iteratively for each run and variable time steps to ensure the program's efficiency, accuracy and stability.

The FRS model *JASMINE* (Analysis of Smoke Movement In Enclosures) uses CFD to describe the heat and mass transfer processes during dispersion of combustion products from a fire. As is generally the case in CFD models, the processes in JASMINE are simulated by the Navier-Stokes equations. The software includes various parameters of mass and heat transfer during the movement of smoke and has been developed, validated and improved over two decades.

JASMINE is considered unique among CFD fire models because of its realistic treatment of the 'fire science' and its extensive validation over a long period in a wide variety of locations like atria, hospitals, sport stadia, warehouses, tunnels, transportation vehicles and ships all with different features. While the model primarily focused on enclosed spaces, it can apparently also be used for modelling smoke propagation in the external environment with different wind and atmospheric conditions.

Outputs can be viewed in vivid graphics and the consequences of a fire simulated to help design fire protection systems or analyse factors for safe evacuation of occupants.

12.8.3 BRANZFIRE

BRANZFIRE is a fire modelling software for a multi-room zone model that can be linked to flame spread and fire growth models. The software can be used to simulate ISO 9705 room corner fires. The model has outputs like hot/cold layer temperatures, vent flows, pressure, room surface temperatures, layer height, visibility, etc.

BRANZ and the University of Canterbury have developed a new model called B-RISK to simulate building fires. The results can be presented in a probabilistic form and is intended to replace the BRANZFIRE program in due course.

Other BRANZ-developed models include TR8 to predict the fire resistance of concrete slabs and floor systems and TR9 to extrapolate the results of a fire resistance test on a light timber framed wall, floor, or ceiling.

12.8.4 SMARTFIRE

SMARTFIRE has been developed by the University of Greenwich as an open architecture integrated CFD code and is currently supported by the same university and the UK Home Office.

The SMARTFIRE project has primary aims of which the first is to make the software an attractive tool for the fire modelling community and the second is to simplify fire field models for users with limited knowledge of CFD practice. Accordingly, SMARTFIRE has two modes of operation viz., the novice and expert modes. In the former mode, most of the control is handled by an expert system component. The latter mode is intended for experienced users who wish to utilise all features of the CFD code and in which all CFD control is available to the experienced user.

Interestingly, the SMARTFIRE project has an associated sub-aim of the project which is to produce a teaching tool which the fire community can use to learn good CFD (and hence fire field modelling) practice.

EXODUS is an egress model developed at the University of Greenwich, UK and tracks the path followed by individuals as they make their way out of the enclosure or are overcome by the fire hazards such as heat and toxic gases. It comprises five interacting sub-models viz.,

- *Occupant*—individuals and all their attributes which are fixed or determined by prevailing conditions.
- *Movement*—outward physical movement of every occupant including pushing, shoving etc.
- *Behaviour*—deals with occupant response subject to attributes.
- *Toxicity*—effects on individual occupsnts.
- *Hazard*—ambient and physical conditions including exits.

12.8.5 Fire Foam

FM Global has developed and maintained FireFOAM, an open source software package. Similar to FDS, FireFOAM is aimed at modelling problems relevant to thermo- and fluid-dynamics and multiphase flow. However, it is specialised in simulating heat and smoke transport in fires.

12.9 THE WAY AHEAD

A majority of the fire models developed are deterministic in nature while personal behaviour (particularly in cases like evacuating a building on fire) are stochastic processes. For instance, the same people in the same premises may act differently on different occasions. The number of people in a building and their mobility are also important. All these put together, terms such as average evacuation time must make way for more appropriate tools and techniques that include probabilities and distributions, besides determining acceptable risk.

This brings up another point. Though much work has been done to calculate ASET (Available Safe Egress Time), arriving at RSET (Required Safe Egress Time) has been difficult as it involves disciplines like toxicology, psychology, sociology, architecture, engineering, mathematics, etc.

The practice and development of fire models have generally focused on learning of fire dynamics, improving design and analysing actual fire scenarios. Efforts to use fire modelling in the course of fire control, however, has been subdued. Effective response to emergencies depends upon the availability of accurate and focused information. The FireGrid consortium led by BRE of UK includes the University of Edinburgh, the London Fire and Emergency Planning Authority and some other organisations was set up to provide an architecture by which the results of computer models of physical fire phenomena can be made available to decision-makers in the context of fire emergencies in the built environment. This involves the modelling of fires in buildings as they actually develop, and uses the high performance computer resources available through the 'Grid'.

The approach is that participating computers used can be located anywhere and each of them would send its output calculations to the original programme, which would then include such inputs to arrive at an overall result for the modelling process. Thus, the time required to complete the entire process would be greatly reduced.

12.10 CONCLUSION

Gigantic strides in the field of computing and information technology in recent years have greatly impacted fire science and fire protection engineering as in all other areas of human activity. In line with these developments, fire models (and particularly CFD models) have turned out to be powerful tools for building designers, fire engineers, fire fighters, fire investigators and many others. In fact, it can be easy to fall into a trap of using them indiscriminately, without a good grip of the two vital areas of fire dynamics and computer modelling. It would also be very useful to recall the quote from the SFPE Handbook at the beginning of this chapter which says that computer models cannot replace experimentation.

It is critical that users of fire models have good base knowledge of fire dynamics and fire protection engineering, besides knowing the capabilities, limitations and appropriateness of the same as well as the assumptions made in any specific model being used. The primary fact is that models supplement engineering judgement and the former is not a substitute for the latter. Further, no **single** run of a CFD or zone model can provide correct answers and it takes a significant amount of work to use such models properly. The incorrect selection and use of a model can have major consequences on diverse fronts such as a smoke control system or escape route intended for life safety, building design factors such as fire resistance of building members and other fire features of passive and active fire protection.

Hence, the following must also be conscientiously observed while using fire models:

- Users must be qualified and experienced fire engineers, competent in both fire science and fire modelling theory.
- Attendance at a short course on fire modelling alone does not make one a competent fire modelling specialist; much practice is required for that.
- Models, sub-models, assumptions must be appropriate for the purpose of use.
- There is no *one answer* to use of a model in a situation and the user must give a range of likely answers based on sensitivity analysis and many runs.
- Regarding the number of runs required for a model, a rule of thumb suggests
  - Zone model, most simple problems: 15–30 runs
  - Zone model, more complicated 30–100 runs
  - Probably fewer, but still many for CFD
- Adequate computer resources and time are required for a modelling exercise to be meaningful.
- Each model has specific limitations as a result of the assumptions made and the experimental methods used to derive correlations and input data.

QUESTIONS FOR REVISION

1. Explain with examples the significance of simulators in aviation and plant operations.
2. What are the zones into which a compartment fire scenario can be divided? Explain the formation of hot and cold layers in that scenario.
3. Laboratory experiments and mathematical models are complementary. Explain.
4. What is the primary difference between the principles of zone and field fire models? Name some instances where CFD models can be used but not the zone models.
5. What is an egress model? Explain some items of scope to which an egress model can be put.
6. Write short notes on some fire models that are accessible on the Internet.

Appendix 1

Explosions Explained

Fire and ***explosion*** are words often used as a pair. Explosions generally refer to a variety of scenarios that result in overpressures but there are some explosions like ***vapour cloud explosion (VCE)*** and ***dust explosion*** that are associated with the process of combustion. Another type of explosion is ***boiling liquid expanding vapour explosion (BLEVE)*** where flammable substances have been involved in many cases and where the genesis and consequence of the scenario involves fire. Given below are brief descriptions of the three phenomena.

A1.1 VAPOUR CLOUD EXPLOSION (VCE)

Liquefied gases (e.g., LPG) and volatile liquids (e.g., naphtha, petrol) can very easily change phase from liquid to vapour if exposed to ambient conditions. The vapours of these products are heavier than air (vapour density = 1), which implies that they tend to settle down close to the ground. Stable and light wind conditions may not be sufficient to disperse the vapours. They can travel down inclines and accumulate in low areas such as trenches and pits. When the quantity of vapour generated is large and it accumulates, such accumulations are known as *vapour clouds*. If the released substance is flammable, its initial concentration will be 'too rich'. In the best situation, prevailing winds and air movements can assist diffusion with atmospheric air till its concentration passes through its flammable range and continues till it becomes 'too lean'. This can also be achieved through human intervention as is done with water sprays and fog nozzles.

If the surrounding atmosphere is relatively stable, diffusion will not be rapid and the cloud will be in the 'too rich' state. Diffusion will commence at the periphery of the cloud which will in due course enter the flammable range. If an ignition source comes in contact with the periphery at this stage, the cloud will burn in that region. Thereafter, the flame front will try to penetrate inward into the inner region of the cloud which is still in the 'too rich' state. The inward travel of the flame front will draw in air into the cloud that will induce quicker diffusion. A loud noise is characteristic of flash burning as the flame travels through a fuel-rich cloud. This is not an explosion but is sometimes erroneously interpreted as one.

If the early encounter between the 'too rich' cloud and an ignition source referred above does not occur, the cloud will continue to diffuse with the surrounding air. With the passage of time, larger portions of the cloud change from 'too rich' into flammable range. The time required for this to happen depends on the properties of the vapour in the cloud, ambient conditions and surrounding constraints such as built structures or natural features such as topography, vegetation, etc. In this time, the size of the cloud will have expanded to a much larger size. To illustrate,

1 kg-mole, i.e., 44 kg propane (C_3H_8) occupies 22.4 kL at STP

$$1 \text{ T propane would occupy } \frac{22.4 \times 10^3}{44} \text{ kL} = 509 \text{ m}^3 \ (1 \text{ kL} = 1 \text{ m}^3)$$

Flammability range of propane in air is 2–10 volume% (approx.)

The combustion reaction of propane with oxygen is

$$C_3H_8 + 5O_2 \longrightarrow 3CO_2 + 4H_2O$$

For 20% (approx.) oxygen concentration in air, for stoichiometric reaction, 1 volume of propane would require 25 volumes of air giving a propane-in-air concentration of

$$\frac{1}{(25+1)} = 3.85\%.$$

As discussed earlier in this text, although a flammable vapour–air mixture can support combustion throughout its flammable range minimum energy for ignition is required at stoichiometric ratio.

Assuming that the cloud is at a uniform 3.85 volume% with air (stoichiometric ratio) concentration, the size of the cloud will be

$$509 \text{ m}^3 \times \frac{100}{3.85} = 1322 \text{ m}^3$$

If the cloud at this stage comes in contact with a source of ignition, very rapid combustion resulting in a violent explosion can occur.

Figure A1.1 Vapourcloud explosion at a petroleum storage terminal.

Serious studies to understand vapour cloud explosions were undertaken from the 1970s after the cyclohexane vapour cloud explosion at Nypro's Flixborough plant in the UK (Figure A1.1). Initially, the phenomenon was described as PUVCE, that is, percussive unconfined vapour cloud explosion. The term ***percussive*** was used as more than one explosion-like sound could sometimes be heard following the primary sound due to the explosive combustion event. ***Unconfined*** was intended to describe that the event occurred outside containment but another school of thought felt that such explosions could not occur in really open areas and that structural or other constraints could act as confinement. In due course, P and U were dropped from PUVCE and today VCE is an accepted nomenclature.

A1.2 BOILING LIQUID EXPANDING VAPOUR EXPLOSION (BLEVE)

The phenomenon of BLEVE is very different from a VCE. BLEVE involves sudden, violent rupture of a container that holds superheated liquid (a vapour held as a liquid above its boiling point by applying pressure, e.g., LPG) and subsequent release of its contents. This release causes sudden drop in pressure from pressurised containment to atmospheric and results in boiling and immediate vaporisation of the liquid at millions of nuclei.

The cause of rupture of the pressurised container may be either due to thermal stresses or due to mechanical stresses, or due to structural fault in the material of construction. However, a majority of BLEVE cases have occurred due to overheating of pressurised containers and the phenomenon is described through the scenario below.

A pressurised container like an LPG bullet has a fire burning beneath it. As the bullet gets heated, there is increased boiling of its contents and this raises the pressure in the vapour space of the bullet. The excess pressure may be vented out through the pressure relief valves and the valves might reseat. This might happen repeatedly with gradual lowering of liquid level and hence wetted surface inside the vessel. As long as the heated portions of the shell is also wetted, the liquid inside will absorb much of the incident heat. With lowered liquid level, flames may impinge directly on 'unwetted' portions of the shell, thus weakening it. Weakening of the shell may also occur around the relief valves if the exiting jets get ignited from some source and we have jet flames. The combined effects of weakened shell and high internal pressures can cause the shell to rupture causing violent release of contents and projecting missiles of container fragments over large distances.

The rupture and sudden release causes a pressure drop inside the vessel resulting in instantaneous boiling at millions of liquid nuclei. These further result in very high increase in volume leading to a massive wave of overpressure and explosion. This form of failure is known as *BLEVE*.

If the contents are flammable like LPG, the sudden vaporisation may lead to formation of a vapour cloud which may get ignited. If ignition occurs when the cloud is in 'too rich' condition, flash fire may occur and a fireball may form.

A BLEVE will normally have missile effects due to violent rupture of the affected container. In addition, in the case of flammable materials like propane, fireballs may be seen and there may be fire propagation on combustible materials in the area. If there are similar storage containers in the path of released missiles, additional containers may also BLEVE (Figure A1.2).

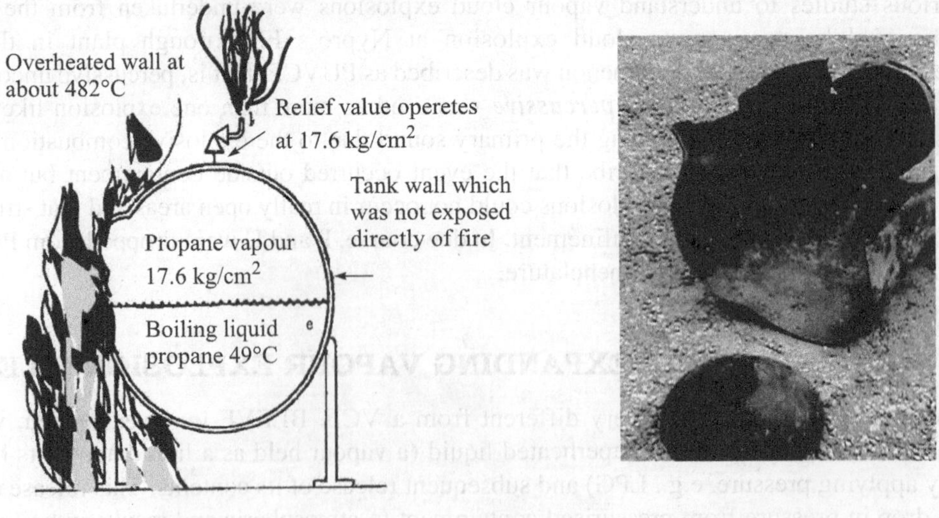

Figure A1.2 BLEVE phenomenon and cylinder that BLEVE'd.

The likelihood of BLEVE on storage vessels can be reduced through design and the concept of mounded storage has been put into practice in large capacity LPG storages.

A1.3 DUST EXPLOSION

Dust explosions have caused deaths and damage on many occasions. Such explosions are caused by the ignition of a suspended mass of combustible dust particles in air in explosive concentration. A wide variety of dusts are known to have exploded under different conditions. These include grain, flour, sugar, plastic, wood, some metal dusts, etc. As flammable gases have a flammable range, dusts have an explosive range. An air–dust mixture containing fine dust is more explosive than a mixture with coarse dust because as the former has greater exposed particle surface. A small dust explosion can shake dust loose and produce another (secondary) dust explosion. Such explosions can occur in a chain and cause massive damage to the premises where it happens.

There are different technical definitions for ***combustible dust:***

- "A solid material composed of distinct particles or pieces, regardless of size, shape, or chemical composition, which presents a fire or deflagration hazard when suspended in air or some other oxidising medium over a range of concentrations". —OSHA, USA
- "A dust that can create an explosive atmosphere when it is suspended in air in ignitable concentrations". —The Alberta Occupational Health and Safety Code in Canada
- NFPA 652-2019 defines a combustible dust as "a finely divided combustible particulate solid that presents a flash fire hazard or explosion hazard when suspended in air or the process-specific oxidizing medium over a range of concentrations."

For ease of understanding, a ***combustible dust*** may be described as a mixture of any fine material with air in concentrations that can catch fire and explode when a source of ignition is applied. Combustible dusts can be from:

- Naturally-occurring organic solids, e.g., coal, sulphur, grains, wood.
- Processed organic solids, e.g., flour, powdered milk, sugar.
- Man-made organics, e.g., pharmaceuticals, pesticides, plastics.
- Metals and alloys, e.g., aluminium, bronze, magnesium, zinc, iron.
- Some non-metallic inorganic materials.

Though some of these materials may normally not be considered 'combustible', they can burn or explode if the dust particle size and its concentration in suspension are both right.

Ignition for the explosion may be either a flame or a spark of sufficient energy produced by friction, static electricity, sparks from an electric appliance, etc.

Dust can accumulate on horizontal or inclined surfaces, ducts, crevices, etc. When the dust is disturbed by any means to form suspensions of explosible concentrations, a serious explosions can occur. The build-up of even a very small amount of dust can cause serious damage. A primary precaution against dust accumulation is housekeeping but brooms and compressed air must not be used; vacuum cleaning is a preferred method of removing dust. Exhaust fans may be used to draw dust into ducts that discharge at a safe location but all machinery must be earthed so that static charges cannot accumulate. Equipment that produce sparks must be shielded to prevent escape of sparks into dusty atmospheres. Any fires (even small ones) on dust must be extinguished immediately or they can escalate to dust explosions.

The Process

As compared to the fire triangle [(1) fuel, (2) ignition source and (3) oxygen], a dust explosion needs two additional elements to make what is known as the 'dust pentagon' (Figure A1.3):

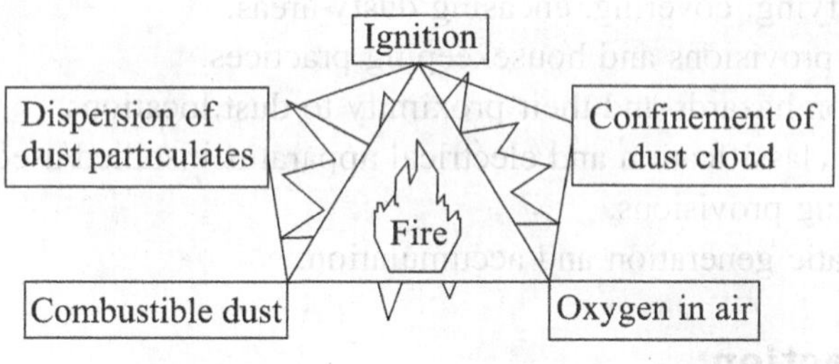

Figure A1.3 Dust explosion pentagon. (*Source:* OSHA).

(4) Dispersion (cloud) of dust particles in the air.
(5) Confinement of the cloud.

Element 4 refers to the suspension of dust particles in the air in the right concentration while 'confinement' in element 5 implies that the dust suspension is in an enclosed or restricted space. This restriction allows pressure build-up and increases the likelihood of explosion. The primary explosion will shake other dust that has accumulated. When this dust becomes airborne, it also ignites. This secondary dust explosion is often more destructive than the primary one.

A number of other conditions must also be met for a dust explosion to occur. These include:

- the dust must be in suspension in air, combustible, and in explosible range.
- an ignition source must come in contact with the dust suspension.
- size of the dust particles must aid flame spread.
- confinement or physical barrier of the dust cloud.
- hest release must be adequate to sustain the fire.
- ventilation and availability of oxygen to sustain combustion.

Containing Explosion Pressures

In general, process equipment design may not cater to withstand pressures likely to be generated by unvented dust explosions as structures required for achieving the required strength will make the equipment very heavy and expensive as also difficult to operate or maintain. Thus, containing explosion effects may not be either feasible or viable. However, a very special case of potential dust explosion could be handling or processing toxic powders. If an explosion does occur here and people present could be exposed to toxic dust, there is no alternative to installing equipment that will not rupture even at maximum expected explosion pressure.

Precautions

It may not be wrong to say that the existence of an event like a dust explosion is not common knowledge. However, such explosions can occur in places that generate or handle small particulates that could form suspensions in the air. Hence, it is advisable to conduct risk assessments and regular inspection of the premises with focus on

- dust generation and accumulation including in hidden areas.
- scope for modifying, covering, encasing dusty areas.
- dust collection provisions and housekeeping practices.
- potential ignition hazards and their proximity to dust locations.
- hazardous area classification and electrical apparatus installed/used.
- explosion venting provisions.
- potential for static generation and accumulation.

Methods of Protection

Industrial equipment like process vessels are normally not designed to withstand the pressures exerted even by minor dust explosions. Hence, designs aim to negate the process of explosion after it has been initiated by preventing the build-up of damaging overpressures in three different ways, viz.,

- Venting
- Automatic suppression
- Containing the explosion in a section/equipment designed to withstand the explosion effects through the use of quick-acting shutoffs and not allowing it to propagate else where.

The choice of one or more of these techniques for implementation will depend on potential explosion pressures, feasibility and viability.

Appendix 2
Remembered for the Wrong Reasons

Some Fire Incidents in the Built Environment in India

The Great London Fire (1666), the Chicago Fire (1871) and Piper Alpha (1988) are among those landmark cases of fire that made people sit up and bring in new regulatory standards or changes in the existing ones. Fire investigation reports are great learning resources for regulators, scientists, engineers, architects, social scientists, psychologists, doctors, lawyers and other professionals. Organisations such as the Chemical Safety and Hazard Investigation Board (CSB) of the US have contributed immensely in opening up investigation findings as their reports are presented in public. Otherwise too, there is a lot of information about major fires that have occurred in the developed world as published reports which have been uploaded on the Internet. Past incidents are revisited to review findings based on newer available knowledge. In India, we have major portions of the M.B. Lal Committee report on the fire at the Jaipur terminal of Indian Oil Corporation Ltd. (2009), accessible on the Internet but this is an exception; fire investigation reports generally are kept 'confidential' thus closing out avenues of potential learning.

Included in this annexure are reports of three very brief reports of fires at different work locations followed by two fire incidents all of which the author investigated. The reports on the Kochi (1984) and Mathura (1988) refineries contain some suggestions for improvement which are standard provisions stipulated by OISD today. The reports were published in Indian and foreign publications and are included here as written soon after the investigations were conducted.

Berthed Cargo Vessel

Sometime in the 1980s, a cargo vessel berthed at Indira Docks of Mumbai harbour reported a fire in one of its holds that contained nylon bales. When the city fire brigade was conducting its fire fighting operations with water jets, the hatch covers of the adjacent hold were violently thrown open. It turned out that the second hold contained calcium carbide in metal drums and when fire fighting water entered the hold, it came into contact with the carbide in some damaged drums. As is well-known, unstable and highly explosive acetylene gas is released when water contacts calcium carbide. Neither the crew of the vessel thought it necessary to inform fire fighters about the cargo on the vessel nor did the fire brigade personnel seek such information.

Foam Rubber Factory

Sometime in 1989–1990, a factory manufacturing foam rubber mattress on the UP-Delhi border suffered a major fire during peak production period. The fire was noticed on a Sunday (weekly off) by factory security personnel. Foam rubber mattresses or cushions need to be cooled after they are formed. This is achieved by spreading the formed products one high. However, due to high production rate, sufficient free space was not available within the factory building and the formed mattresses were piled several high. The heat from the products was thus trapped within the piles. This resulted in spontaneous heating leading to ignition of the mattresses. The site was not insured for spontaneous combustion.

Sensitive Equipment Factory

The issue section of a sensitive product manufacturing organisation in Ghaziabad suffered a major fire on a cold January morning in the early 1990s. The factory worked three shifts round the clock. It was reported that the fire originated when a burning fluorescent light fell from the ceiling to the ground. A closer examination showed a prominent V burn mark on the lower side of a wooden office table. It is very likely that the occupant of the table had a room heater on the floor kept near his feet which went on to ignite the wooden table and other combustible matter in the vicinity.

A2.1 THE COCHIN (NOW KOCHI) REFINERIES FIRE (1984)

Most refinery fires are extinguished in their early stages and not many people outside the refinery come to know about them. The Cochin refinery explosion and fire that took place in the wee hours of 8 March, 1984, however, literally shook many people out of their sleep.

The incident has been variously described by different news reports and magazines. Most of them described it as an explosion in aviation turbine fuel (ATF) tank number 290 and that this was followed by a devastating fire. This report is based on observations made on site nearly a month after the incident.

Refinery

Cochin Refineries Ltd. (CRL) is a joint venture of the Government of India and the Phillips Petroleum Company of USA. The refinery was commissioned as a 2.75 MTPA (Million Tonnes per Annum) fuel refinery in 1966 and subsequently expanded to 3.3 MTPA. It was due for further expansion to 4.5 MTPA and the refinery was scheduled to be shut down in mid-March to hook up the expansion facilities with the existing units.

The design of the refinery has been associated with compactness. Compactness has facilitated operation and kept energy losses low. Manpower requirements have also been kept low by means such as not having a full time fire fighting force. To handle emergencies such as fire, therefore, personnel from various refinery functional areas were trained to man appliances and act as a fire fighting team.

Fire

At 0542 hours on 8 March, 1984, process operators reportedly heard an explosion. They came out of the control room and saw some of the storage tanks south of the unit on fire. The explosion was heard and felt in the refinery housing colony as well as in areas some 2 km away. Refinery personnel who went to the site immediately report that three tanks viz., tanks 30 and 40 containing raw naphtha and tank 290 containing JP-5 (aviation fuel) were involved in a fire.

The fire seems to have started big and the available CRL fire tender stationed in the fire equipment shed (35 m from tank 290) was damaged in the explosion. Together, these factors prevented refinery personnel from actively fighting the fire. In fact, fire fighting operations began only after the appliances from the Cochin Fire Station of the Kerala State Fire Service (KSFS), 13 km away, arrived shortly before 0600 hours (A CRL spokes person mentioned that the first KSFS tender reached the spot at 0625 hours).

Appliances from other KSFS stations up to 77 km away and those belonging to FACT (Fertilizers and Chemicals Travancore Ltd.), the Indian Navy and Cochin shipyard fought the fire. Altogether 16 tenders and several pumps were used in the fire fighting operations. Monitors located in accessible places were also used.

The fire fighting operations were made more difficult because naphtha from tank 30 had spilled through a damaged pipeline into the dyke and there was fire in the dyke too.

The LSHS tank 190 had also developed a leak in the bottom and its contents were spilling into the dyke. As cooling water to the tanks was being applied in larger quantities than could be drained out, water levels in dykes continued to rise. In fact pumps had to be used to dewater the dykes.

Radiant heat from the fire kept fire fighters at bay and their water jets were not fully effective in cooling the tanks. Foam applied to the dyke was disturbed by the water jets allowing reignition of open hydrocarbon surfaces within the dyke.

Fire in the dyke was put out by 0830 hours and that in the chemical warehouse by 0900 hours. Fire in tank 290 had died out by noon and cooling of tanks 30 and 40 was continued thereafter. Around 1800 hours, however, a strong wind blew over the tanks towards the unit. At the same time, LSHS spillage in the dyke from tank 190 caught fire. The arrival of the crash tender from the airport at this crucial time was very helpful and after 2000 hours, fire was again confined to tanks 30 and 40.

These tanks were allowed to burn under controlled conditions till fuel was exhausted around 1030 hours on March 11.

Cooling operations from the direction of the LPG plant was kept up all through to prevent the spread of fire in that direction.

Sixty thousand litres of foam compound were used and some more had to be brought in from outside the refinery.

Damage

The damage to property inside the refinery was spread over a large area. The utilities area, the chemical warehouse and storage tanks 30, 40 and 290 were most affected by the explosion and fire.

The utilities area had very little fire damage but evidence of extensive damage due to explosion was noticeable. The cooling tower was totally destroyed and water headers extending to it remained hanging in mid-air. The turbo generator house was structurally damaged and its corrugated asbestos cement sheet roof was totally destroyed. Structures housing the electric load centres (control and distribution panels) were almost totally destroyed.

The boiler control room which had a near-total glass wall on the unit (west) side was destroyed and instruments on the control panel were damaged to various degrees. Two clocks located in the control room were damaged with their hands indicated 5:42 and 5:43, respectively. The explosion apparently occurred at that time. Some papers in the control room bore burn marks.

There were no visible signs of damage in the process unit area when we visited the site except for refractory damage in a process heater. Any other damage that may have occurred must have been repaired.

The chemical warehouse was a mass of rubble after the fire and explosion. Most of the drums stored in the warehouse were damaged and many of them contained catalysts which may have to be rejected. Some of the equipments for the expansion programme which were incidentally kept near the warehouse were damaged.

Tank 290 on JP-5 duty and tanks 30 and 40 containing naphtha were completely damaged.

The heat due to the fire and the cooling water applied together led to the collapse of these tanks. Some other tanks in the area containing LSHS, furnace oil and long residuum were marginally damaged.

Glass panes in the administrative building and the canteen beyond it were broken in the explosion. False ceilings in these buildings were also affected. The fire equipment shed and the dispensary near the main gate were partially damaged. Tile-roof tenements on the south of the refinery and a church in the area suffered varying degrees of damage.

Two persons (a CRL employee and a contractor's worker) were killed on the spot and two others died later in hospital. Apparently, the post-mortem findings attributed death as due to burns and not due to shock.

Loss due to the fire and explosion is estimated around ₹ 12 crores.

Probable Cause

From the pattern of damage, it appears that there had been an explosion in the utilities area. This is borne out by the fact that there was very little fire damage/burn marks in the area, while structures and equipment were severely damaged.

There were no hydrocarbon pipelines in the utilities area except for fuel lines to the boilers. For the explosion that took place, there had to be an accumulation of fuel vapours and formation of flammable fuel–air mixture. As for sources of ignition, there could be many in the area: boilers, process heaters, electrical systems, hot surfaces, frictional sparks, etc.

Water to the cooling towers is process-return water from various heat exchangers (coolers). The pressure on the water side of the coolers being less than that on the hydrocarbon side, any internal leak will result in hydrocarbon entering the water stream. If the hydrocarbon involved is unstabilised naphtha or LPG, flammable vapours can be released at the cooling tower.

Studies have indicated[1] that for even a 1 in 1000 chance of an unconfined vapour cloud explosion (UVCE), a tonne of flammable vapour is required. For an LPG production rate of

0.25 t/hr, a 1% leak will require 400 hours for a vapour cloud of 1 T to form. Over such a long time span, any flammable vapour leak will be dispersed into the atmosphere before it can form a cloud. Any larger leak will be discovered by plant material balance.

Following the incident, the CRL management had apparently carried out a check and observed a leak in an LPG cooler. It is, therefore, probable that LPG had escaped with the return process water to the cooling tower. However, as discussed above, this leakage alone could not have provided fuel for the entire cloud and there possibly was some other source of fuel too.

Observations

Setting up and operating a refinery involves big expenditure and so cutting costs wherever possible is important. Unfortunately, this often leads to neglect of safety considerations. A compact plant saves on materials, energy and manpower and, like many other refineries, the one at Cochin was designed that way.

Fired process heaters are bunched together and close to other process equipment. The process and utilities sections are separated only by a lane. The extreme is of course the location of a warehouse for chemicals in the space between the utilities section and storage tanks. It might be argued that mandatory spacing requirements were in fact observed but this should be done in combination with layout considerations. The designers of the project obviously did not take such a comprehensive view.

The refinery was commissioned in 1966 and will, therefore, reflect the design philosophy of the early sixties. This may be the reason for the control rooms having glass doors and panels on the side facing the unit. Besides, doors of the control room open towards the unit and, in the even to an emergency, plant personnel will not have an escape route leading to the open.

The storage tanks located south of the chemicals warehouse had dykes barely enough for the tank volumes. The separation distances between the tanks just about satisfied the requirements of the Petroleum Rules (1976). The water monitors located in the tank farm could not be effectively directed to reach all the tanks. The tanks had no fixed foam protection system, and temporary arrangements for foam application had to be made during the fire. Hoses employed in these arrangements were burnt in the fire.

The storage tanks and the LPG plant are located close to the south wall of the refinery, just beyond which are a public road and tenements. Property on either side of the wall is, therefore, exposed to hazards on the other side.

The chemical warehouse was constructed of laterite bricks, steel frame and asbestos cement sheet roof. Catalysts in drums, water treatment chemicals and a host of other hydrocarbon-based materials were stocked in this warehouse. As there were warehouses of better construction near the administrative building, it appears that the chemical warehouse had come up as a temporary measure but continued to be in use.

The Petroleum Rules (1976) require that every refinery has a 'well-organised and trained fire fighting service'. It is not clear whether CRL's fire fighting set-up satisfies this requirement.

It may also be noted that refinery plant and equipment are located closer to the gate than in most other refineries. The administrative office, by virtue of its location within battery limits, is exposed to the hazards from the plants and facilities.

It is said that what has happened can happen again. The 8th March incident in CRL can occur there or elsewhere if proper precautions are not taken. Planning, design, operation and maintenance functions must use engineering judgement or seek expert opinions to identify and evaluate hazards; satisfying statutes may not be enough.

It may also be appropriate to take a critical look at mandatory requirements and suggest improvements. For example, rules regarding provision of private fire brigades can be more specific. Rules should also give bases for requirements such that their application does not appear arbitrary.

Cochin and its neighbouring areas are becoming a centre for industrial investment and the meagre public fire brigade facilities must be improved. Large industries of the area should also form a plan for mutual aid to handle fire emergencies.

CRL must review its fire protection arrangements. Dyke walls around tanks can be raised to increase the capacities of the dykes. Monitor facilities can be increased particularly on dykes around the tanks. Fixed foam protection to tanks must be provided.

Discussions on the CRL experience should be open because it is in everyone's interest that such incidents do not recur.

A2.2 MAJOR REFINERY FIRE

For personnel who worked in the Indian Oil Corporation refinery at Mathura in the state of Uttar Pradesh, August 30, 1988, was just another day. But in the next few hours they were to experience the worst disaster in the refinery's history.

Around mid-morning that day more than 40 tank wagons (each of 27 kL capacity) containing naphtha were involved in a fire which claimed six lives. Loading facilities were severely damaged as were the overground hydrant lines and the railway tracks.

The Mathura refinery complex, which is located on the Delhi–Agra highway, 15 km south of Mathura city, was commissioned in 1982 as a 6 MMTPA refinery. Its capacity today *(at the time of the incident and writing this report)* is 7.5 MMTPA, making it the second largest refinery in India.

Crude oil for the refinery is received from Bombay High and imported through a pipeline from Salaiya in the State of Gujarat. Products from the refinery are despatched to the northern states of India in rail tank wagons and road tankers.

Filling the tank wagons was done by using three broad gauge and one metre gauge railway gantries located towards the northern periphery of the refinery and north of the product tank farm and the quality control laboratory.

Each gantry has facilities to load on either side of it. There is a separation of approximately 15 m between the adjacent tracks of two gantries, which serves as a graveled road for movement of vehicles. The gantry sheds themselves were built of exposed steel members with corrugated asbestos (AC) sheet roof.

A gully sucker used for recovering oil spillages from drains and sumps had broken down in the area between gantries BG1 and BG2 and a jeep carrying two refinery employees was proceeding to the spot when the incident occurred.

Refinery personnel in other sections noticed smoke in the gantry area and informed the Executive Director and the General Manager of the refinery. The Manager (Fire and Safety) was in the main administrative building outside the plant battery limits and heard a loud noise around the same time.

All these officials went immediately to the site. At the time of the incident, there were 67 and 75 wagons respectively in gantries BG1 and BG2; some had been filled and others were waiting to be filled.

Fire fighting operations were started by the refinery's fire brigade with its three foam and one dry chemical powder tenders, with water from the hydrant system. The fire crew of 13 men found it difficult, however, to combat the fire and brigades from Mathura, Bharatpur and Agra were called at 1100 hours. In all, 20 fire tenders were engaged to control the fire which was fully extinguished at 1400 hours.

A total of 20 kL of foam compound and 2 tonnes dry chemical powder was used to combat the fire, although their method of application is not known. The operations also involved detaching 96 wagons from the affected rakes and moving them to a safer place.

Gantries BG1 and BG2 (where wagon filling was in progress) suffered heavy damage and nearly half of each was almost totally destroyed. A third gantry was idle and damage was only in some sections. Naphtha in the damaged tank wagons and pipelines was lost in the fire, though the exact quantity is not known.

Six Lives Lost

At ground level, overground fire hydrant lines, product lines and railway tracks were deformed, while the wagons and buffer springs had been damaged by heat. Combustible parts of the jeep and the gully sucker such as tyres, upholstery and rubber hoses had been burnt and their metal parts damaged to various degrees. Foliage on trees south of the gantry were charred but windows on the laboratory remained intact.

Unfortunately, the greatest loss in the incident was six lives; the two occupants of the jeep and four contractor's men working on the gully sucker.

Normally, product vapours will be present in the gantry area while filling wagons with volatile products. Fortunately, generally prevailing wind conditions can be expected to disperse the vapours rendering the vapours 'too lean' to burn. In the case of a sustained release of liquid due to such causes as fracture of product line or overfilling of wagons, however, efficient dispersal may not be effected and a vapour cloud may be formed.

Pikaar[2] states that liquid on the ground or water will continue to boil if its atmospheric boiling point is below ambient temperature. If the boiling point is above ambient, vaporisation and cloud formation is brought about by the passage of air over the liquid surface.

While in the second case, cloud formation will be minor, the downward flow of liquid along a structure such as a tank wall will promote cloud formation. The dense vapour–air mixture will sag to the ground and allow fresh air to meet the liquid and promote the evaporation process.

Following a release, vapour will be distributed in various directions and there may be several zones where the fuel–air mixture is within the flammability range. When such a zone encounters an ignition source, local ignition will occur and then fire may spread throughout the cloud.

Spillages in the gantry area of Mathura Refinery do not appear to be unusual as evidenced by the presence of the gully sucker. However, on the day of the incident there could have been a sustained overflow of naphtha from one or more of the tank wagons resulting in excessive spillage beneath the wagons and in the formation of a flammable vapour cloud.

Considering that there is a separation of only about 15 m between the tracks and that there were rakes on both of them, it is likely that there was little dispersal or lateral movement of the cloud and that it was stagnant while growing in size.

As explained earlier, such a cloud could be a pre-mixed fuel–air mixture. Once ignited at any of the flammable zones, the flame could travel through the cloud and then become several separate fires at the points from where liquid fuel is generating vapours. Thus, in the present case, the cloud was probably ignited at one point and this was followed by fires in the drains, beneath the wagons and also at the filling points on top of the wagons.

Electrical equipment installed in the gantry area is of the flameproof type, there is a ban on smoking (apparently observed) in the area and no hot work was being carried out on that day. The weather was clear and there was no cloudburst or lightning strike.

Static charges developed during filling operations could have ignited vapours at the wagon top but, in all probability, flames would have been confined to that point until the wagon ruptured. The gully sucker was idle and could not have acted as the ignition source.

The jeep carrying the refinery personnel did not have a spark arrestor on its exhaust, and sparks discharged through the exhaust could have ignited the cloud as the jeep came into its vicinity. Another potential ignition source could have been an electrical fault in the engine area of the jeep. As regards possible ignition by hot metal surfaces of the engine or exhaust pipe, the likelihood is low.

This is inferred from the American Petroleum Institute (API) publication, Ignition Risk of Hot Surfaces in Open Air[3] ignition by a hot surface should not be assumed which states that ignition by a hot surface should not be assumed unless the surface temperature of a hot surface is at least 200°C above the normal accepted minimum ignition temperature. For gasoline (close to naphtha), ignition temperature is 280–425°C and the hot surface temperature without ignition is 540–725°C (API).

It was learnt from refinery personnel that three to four wagons are filled simultaneously at the gantries and that filling takes nearly three hours. Thus the filling rate could be 27 kilolitres an hour or more, and an overflow for even ten minutes would imply a spillage of 4.5 kL and flowing down the wagon surface enhancing premixed fuel–air cloud formation.

Even if the overflow had been stopped, the cloud may not have been efficiently dispersed and it existed in the area between the rakes when the jeep arrived. An electrical fault on the engine side of the jeep or its exhaust ignited the cloud.

It is not essential that flame travel will take place only if the cloud is within the flammable limits as a flame will progress even along a fuel-rich cloud. Only the flame speed will be lower, burning will be luminous and over pressures may not be created; there will also be intense thermal radiation.

It is reported that there was a loud noise and this is a characteristic of flash burning as the flame travels through a fuel-rich cloud. The charring of foliage on the trees near the laboratory building indicates the intensity of thermal radiation; there was no damage to the window panes of the building, indicating the absence of overpressures. Thus, there seems to have been a flash fire.

As the flame travelled through the cloud, it ignited liquid pools on its path and also started fires at the filling points of the wagons. Thus there were several fires which could have burned simultaneously — at the jeep, on spillages and on top of the wagons. The last of these damaged the filling lines releasing more fuel and the ground fires damaged ground-level product lines worsening the situation in the same manner. The release of smoke from the fires could make it difficult to distinguish the fires and it could appear as one, making fire-fighting operations difficult.

Pool fire flames could have been twice as high as the pool diameters and thermal radiation effects are high, and this could explain the damage to lower portions of the wagons. Exposed fire hydrant lines and the railway tracks did not find the scope for lateral expansion when heated by fires and were distorted. Exposed structural members of the gantry sheds buckled and collapsed due to the thermal stresses and the sheet roofs disintegrated both from heat and from falling down.

Flame travel speed in flash fires could have been as high as 20 m/s and the persons in the jeep could have been trapped inside a burning zone. The others were probably affected by thermal radiation causing intense burns.

While it may be tempting to recommend additional mechanisms to ensure safety, it is advisable and easier to improve on work practices. Some of these are:

- An operator must be present while filling is in progress and the wagons must never be left unattended.
- If it is essential for the operator to be temporarily absent, a relief should attend the wagons.
- Dipping must be done more frequently once the wagon is half-filled.
- Pumping rate should be limited to $vd \leq 0.5$ where v (m/s) is the flow velocity and d(m) is the pipe diameter, and linear velocity must never exceed 7 m/s.
- No automobiles should be allowed to enter the area when volatile products are being filled. Emergency vehicles such as fire tenders or ambulances which may be required to attend must have spark arresters and their electrical systems should be in perfect order. The movement of ordinary diesel locomotives should also be restricted.
- As the filling pumps are operated by refinery personnel and the wagons are attended by marketing division employees, there is scope for communication gaps. Person-to-person meetings must be held at the start of filling operations.
- Extra care must be taken while 'switch loading', as the filling of intermediate vapour pressure products into wagons that previously contained high vapour pressure products has caused many fires.

Besides those recommendations, the following modifications are also suggested:

- Fill pipes should go down to the bottom to avoid splash filling. Linear velocity of liquid flow should not exceed 1 m/s till the end of the pipe is immersed.
- Steel columns of the gantry should be encased in concrete or encased otherwise (such as by spraying gypsum slurry plaster or by applying vermiculite cement). Roof trusses should be treated with fire-resistant paints.
- Fire hydrant lines should be laid in trenches, covered with removable concrete slabs.
- Tower-mounted monitors at the east and west ends of the yard could help combat fires in the congested gantry area.

- A water spray system may be considered for installation all along the gantries to provide protection to the rakes.
- Sufficient gradient should be provided in the gantry yard drains and directed to one or more sumps on the outer periphery of the yard. A gully sucker then would not need to enter into the filling area in normal situations.

Hindsight provides us with a tool to point out what should not have been done; the tool is useful if such acts and incidents are not allowed to recur. But there are acts of omission, which creep into daily work through complacence and it is surprising that even knowledgeable people allow those to happen. The entry of the jeep without a spark arrester is one such example.

Such incidents should be highlighted to prevent further disasters.

REFERENCES

1. Kletz, T.A., *Referred in Loss Prevention in the Process Industries*, 1980, p. 583.
2. Pikar, M.J., Unconfined Vapour Cloud Dispersion and Combustion — An Overview of Theory and Experiments, C20-C34. In: 4th International Symposium on Loss Prevention and Safety Promotion in the Process Industries, Sept. 1983. IChE.
3. Ignition Risk of Hot Surfaces in Open Air, API PSD 2216, American Petroleum Institute.

Appendix 3

Indian Statutes/Standards Related to Fire Safety

A3.1 SELECTED STATUTES/GOVERNMENT GUIDELINES

(a) Model building by-laws, Ministry of Urban Development, Government of India
(b) Building by-laws of respective cities
(c) The Factories Act (1948) with ammendments
(d) Explosives Act (1884) and Rules (2008)
(e) Petroleum Act (1934) and Rules (2002)
(f) The Mines Act (1952)
(g) The Oil Mine Regulations (1984)
(h) Disaster Management Act (2005)
(i) Fire Services Act and Fire Services Rules (of States)
(j) The Delhi Fire Prevention and Fire Safety Act (1986) and Rules (1987)
(k) The Maharashtra Fire Prevention and Life Safety Measure Act (2006) and Rules (2009)
(l) Manufacture, Storage and Import of Hazardous Chemicals (MSIHC) Rules, 1989
(m) Guidelines on Incident Response System published by the National Disaster Management Authority
(n) Guidelines on Formulation of State DM Plans, published by the National Disaster Management Authority
(o) Guidelines on scaling, type of equipment and training of fire services, published by the National Disaster Management Authority
(p) Chemical Accidents (Emergency Planning, Preparedness and Response) Rules, 1996
(q) The Petroleum and Natural Gas Regulatory Board (Codes of Practices for Emergency Response and Disaster Management Plan (ERDMP)) Regulations, 2009.

A3.2 BUREAU OF INDIAN STANDARDS

Sl. No.	*IS*	*Title*
1	884-1985	Specification for first-aid hose-reel for fire fighting
2	901-1988	Specification for couplings, double male and double female instantaneous pattern for fire fighting
3	902-1992	Specification for suction hose couplings for fire-fighting purposes
4	903-1993	Specification for fire hose delivery couplings, branch pipe, nozzles and nozzle spanner
5	904-1983	Specification for 2-way and 3-way suction collecting heads for fire-fighting purposes
6	905-1980	Specification for delivery breechings, dividing and collecting, instantaneous pattern for fire-fighting purposes
7	906-1988	Specification for revolving branch pipe for fire fighting
8	907-1984	Specification for suction strainers, cylindrical type for fire-fighting purposes
9	908-1975	Specification for fire hydrant, stand post type
10	910-1980	Specification for combined key for hydrant, hydrant cover and lower valve
11	926-1985	Specification for fireman's axe
12	927-1981	Specification for fire hooks
13	928-1984	Specification for fire bells
14	937-1981	Specification for washers for water fittings for fire-fighting purposes
15	939-1977	Specification for snatch block for use with fibre rope for fire brigade use
16	941-1985	Specification for blower and exhauster for fire fighting
17	942-1982	Functional requirements for 275 l/min portable pump set for fire fighting
18	943-1979	Functional requirements for 680 l/min trailer pump for fire brigade use
19	944-1979	Functional requirements for 1800 l/min trailer pump for fire brigade use
20	947-1985	Functional requirements for towing tender for trailer pump for fire brigade use
21	948-1983	Functional requirements for water tender type 'A' for fire brigade use
22	949-2012	Functional requirements for emergency rescue tender for fire brigade use
23	950-2012	Functional requirements for water tender type 'B' for fire brigade use
24	951-2003	Functional requirements for crash fire tender for air fields
25	952-1986	Specification for fog nozzle for fire brigade use
26	955-1980	Functional requirements for dry powder tender fire brigade use
27	956-2004	Functional requirements for rescue tender for air fields
28	957-1967	Specification for control van for fire brigade use
29	1641-1988	Code of practice for fire safety of buildings (general): General principles of fire grading and classification
30	1642-1989	Code of practice for fire safety of buildings (general): Details of construction
31	1643-1988	Code of practice for fire safety of buildings (general): Exposure hazard
32	1644-1988	Code of practice for fire safety of buildings (general): Exit requirements and personal hazard

Sl. No.	*IS*	*Title*
33	1646-1997	Code of practice for fire safety of buildings (general): Electrical installations
34	1941 (Part 1)-1976	Functional requirements for electric motor sirens: Part 1 AC, 3 Phase 50 hz, 415 V type
35	2097-2012	Specification for foam-making branch pipe
36	2175-1998	Specification for heat sensitive fire detectors for use in automatic fire alarm systems
37	2189-2008/ ISO/TR 7240-14	Code of practice for selection, installation and maintenance of automatic fire detection and alarm system
38	2190-2010	Code of practice for selection, installation and maintenance of first-aid fire extinguishers
39	2546-1974	Specification for galvanized mild steel fire bucket
40	2696-1974	Functional requirements for 1125 l/min light fire engine
41	2726-1988	Code of practice for fire safety of industrial buildings — Cotton ginning and pressing (including cotton seed delintering) factories
42	2745-1983	Specification for non-metal helmet for firemen and civil defence personnel
43	2871-2012	Specification for branch pipe, universal for fire-fighting purposes
44	2878-2004	Specification for fire extinguisher, carbon dioxide type (portable and trolley mounted)
45	2925-1984/ ISO 3873	Specification for industrial safety helmets
46	2930-1980	Functional requirements for hose laying tender for fire brigade use
47	3034-1993	Code of practice for fire safety of industrial buildings: Electrical generating and distributing stations
48	3058-1990	Code of practice for fire safety of industrial buildings: Viscose rayon yarn and/or staple fibre plants
49	3079-1990	Code of practice for fire safety of industrial buildings: Cotton textile mills
50	3582-1984	Specification for basket strainer for fire-fighting purposes (cylindrical type)
51	3594-1991	Code of practice for fire safety of industrial buildings: General storage and warehousing including cold storages
52	3595-2002	Code of practice for fire safety of industrial buildings: Coal pulverisers and associated equipments
53	3614 (Part 1)-1966	Specification for fire check doors: Part 1 Plate metal covered and rolling type
54	3614 (Part 2)-1992	Specification for metallic and non-metallic fire check doors: Resistance test and performance criteria
55	3744-1985	Specification for hose binding machine
56	3808-1979	Method of test for non-combustibility of building materials
57	3809-1979	Fire resistance test of structures
58	3836-2000	Code of practice for fire safety of industrial buildings: Jute mills

(Contd.)

Sl. No.	*IS*	*Title*
59	3844-1989	Code of practice for installation and maintenance of internal fire hydrants and hose reels in premises
60	4226-1988	Code of practice for fire safety of industrial buildings: Aluminium/magnesium powder factories
61	4308-2003	Specification for dry chemical powder for fighting B and C class fires
62	4571-1977	Specification for aluminium extension ladders for fire brigade use
63	4643-1984	Specification for suction wrenches for fire brigade use
64	4861-1984	Specification for dry powder for fighting fires in burning metals
65	4862 (Part I)-1986	Specification for portable fire extinguisher for aircraft part 1 halon 1211 type
66	4886-1991	Code of practice for fire safety of industrial buildings: Tea factories
67	4927-1992	Specification for unlined flax canvas hose for fire fighting
68	4928-1986	Specification for delivery valve for centrifugal pump outlets
69	4947-2006	Specification for gas cartridges for use in fire extinguishers
70	4989-2006	Specification for foam concentrate for producing mechanical foam for fire fighting [Amalgamating Part 1, Part 2 and Part 3]
71	4989 (Part 4)-2003	Specification for multipurpose aqueous film forming foam liquid concentrate for extinguishing hydrocarbon and polar solvent fires
72	5131-2002	Specification for dividing breeching with control for fire brigade use
73	5290-1993	Specification for landing valves
74	5486-1985	Specification for quick release knife
75	5505-1985	Specification for multi-edged rescue axe (non-wedging)
76	5612 (Part 1)-1977	Specification for hose clamps and hose bandages for fire brigade use: Part 1 Hose clamps
77	5612 (Part 2)-1977	Specification for hose clamps and hose bandages for fire brigade use: Part 1 Hose bandages
78	5714-1981	Specification for hydrant, stand pipe for fire fighting
79	5888-1970	Code of practice for design and construction of fire service drill-tower
80	5896 (Part 1)	Code of practice for selection, operation and maintenance of special fire fighting appliances: Part 1 Combined foam and CO_2 crash tender
81	5896 (Part 3)	Code of practice for selection, operation and maintenance of special fire fighting appliances: Part 3 Turntable ladders
82	6026-1985	Specification for hand-operated sirens
83	6067-1983	Functional requirements for water tender type 'X' for fire brigade use
84	6070-1983	Code of practice for selection, operation and maintenance of trailer fire pumps, portable pumps, water tenders and motor fire engines
85	6329-2000	Code of practice for fire safety of industrial buildings: Saw mills and wood works
86	7673-2004/ ISO 8421-1	Glossary of terms for fire-fighting equipment
87	7692-1993	Specification of head forms for testing of helmets

Sl. No.	IS	Title
88	8090-1976	Specification for couplings, branch pipe, nozzle, used in hose reel tubing for fire fighting
89	8096-1992	Specification for fire beaters
90	8149-1994	Functional requirements for twin CO_2 fire extinguisher (trolley mounted)
91	8423-1994	Specification for controlled percolating hose for fire fighting
92	8442-2008	Specification for stand post type water and foam monitor for fire fighting
93	8757-1999	Glossary of terms associated with fire safety
94	8758-1993	Recommendations for fire precautionary measures in the construction of temporary structures and pandals
95	9109-2000	Code of practice for fire safety of industrial buildings: Paint and varnish factories
96	9668-1990	Code of practice for provision and maintenance of water supplies and fire fighting
97	9972-2002	Specification for automatic sprinkler heads for fire protection service
98	9995-1981	Specification for visor for non-metal police and fireman's helmet
99	10460-1983	Functional requirements for small foam tender for fire brigade
100	10658-1999	Specification for higher capacity dry powder fire extinguisher (trolley mounted)
101	10993-1984	Functional requirements for 2000 kg dry powder tender for fire brigade use
102	11101-1984	Specification for extended branch pipe for fire brigade use
103	11108-1984	Specification for portable fire extinguishers halon-1211 type
104	11360-1985	Specification for smoke detectors for automatic electrical fire alarm system
105	11457 (Part 1)-1985	Code of practice for fire safety of chemical industries: Part 1 Rubber and plastic
106	11460-1985	Code of practice for fire safety of libraries and archives
107	11833-1986	Specification for dry powder fire extinguisher for metal fires
108	12349-1998/ ISO 6309	Fire protection – Safety signs
109	12407-1988	Graphic symbols for fire protection symbols
110	12456-2004	Code of practice for fire safety of electronic data processing installation
111	12458-1988	Method of fire resistance test of fire stops
112	12459-1988	Code of practice for fire safety in cable runs
113	12717-1989	Functional requirements of fire-fighting equipment – High capacity portable pumpset (1100-1600 l/min)
114	12777-1989	Method for classification of flame spread of products
115	12796-1989	Specification for fire rake
116	12805-1989	Tool Mcleod for fighting forest fires
117	12835 (Part 1)- 1989/ ISO 7203-1	Code of practice for design and installation of fixed foam fire extinguishing system: Part 1 Low expansion foam
118	12853-1989	Specification for pulaski tool for forest fires.

(Contd.)

Sl. No.	IS	Title
119	12859-1989	Specification for brush hook for forest fires.
120	12861-1989	Specification for double bit axe for forest fires.
121	12862-1989	Specification for fire-fighting shovel for forest fires.
122	13039-1991	Code of practice for external hydrant systems provision and maintenance
123	13045-1991	Code of practice for fire safety of industrial buildings: Flour mills
124	13385-1992	Specification for fire extinguisher – 50 litres capacity wheel mounted water type (Gas Cartridge)
125	13386-1992	Specification for fire extinguisher – 50 litres capacity mechanical foam type
126	13694-1993	Code of practice for fire safety in iron and steel industries
127	13716-1993	Code of practice for fire safety of hotels
128	14435-1997	Code of practice for fire safety in educational institutions
129	14609-199/ ISO 7202	Specification for dry chemical powder for fighting ABC Class fires
130	14689-1999	Code of practice for fire safety of industrial buildings (printing and publishing industry)
131	14850-2000	Code of practice for fire safety of museums
132	14851-2000	Code of practice for maintenance of fire hose
133	14951-2001	Specification for fire extinguisher, 135 litres capacity mechanical foam type
134	14993-2001	Specification for high-pressure fire fighting hose
135	15051-2001	Specification for high-pressure fire hose delivery couplings
136	15105-2002/ ISO 6182-1	Specification for design and installation of fixed automatic sprinkler fire extinguishment systems
137	15220-2002/ ISO 7201-1	Specification for halon 1211 and halon 1301 – halogenated hydrocarbon – fire extinguishing media for fire protection
138	15221-2002/ ISO 7201-2	Code of practice for safe handling and transfer procedure of halon 1211 and halon 1301 – (halogenated hydrocarbon) – Fire extinguishing media
139	15222-2002/ ISO 5923	Specification for carbon dioxide as fire extinguishing media for fire protection
140	15301-2003	Code of practice for installation and maintenance of fire fighting pump
141	15325-2003	Code of practice for design and installation of fixed automatic high and medium velocity water spray system
142	15381-2003	Specification of fire blankets
143	15394-2003	Code of practice for fire safety in petroleum refineries and fertilizer plants
144	15493-2004/ ISO14520-1	General requirements for gaseous fire extinguishing systems
145	15496-2004	Code of practice for inspection and maintenance of gaseous fire extinguishing systems
146	15497-2004/ ISO14520-12	Gaseous fire extinguishing systems – IG 01 extinguishing systems

Sl. No.	IS	Title
147	15501-2004/ ISO14520-15	Gaseous fire extinguishing systems – IG 541 extinguishing systems
148	15505-2004/ ISO14520-6	Gaseous fire extinguishing systems – HCFC Blend A extinguishing systems
149	15506-2004/ ISO14520-14	Gaseous fire extinguishing systems – IG 55 extinguishing systems
150	155517-2004/ ISO14520-9	Gaseous fire extinguishing systems – HFC 227ea (Hepta Fluoro Propane) extinguishing systems
151	15519-2004/	Code of practice for water mist fire protection systems – System design, installation and commissioning
152	15525-2004/ ISO14520-13	Gaseous fire extinguishing systems – IG 100 extinguishing systems
153	15528-2004/ ISO 6153	Gaseous fire extinguishing systems – Carbon dioxide total flooding and local application (Sub-floor and in-cabinet), high and low pressure (refrigerated) systems
154	15682-2006	Fire fighting vehicle and equipment – Symbols for operator control and other display
155	15683-2006	Portable fire extinguishers – Performance and construction – specification
156	15811-2008	Specification for long range water-cum-foam monitor with self-inducting, monitoring and non-aspirating jet and spray/fog nozzle for fire fighting
157	15821-2006	Gaseous fire extinguishing system – Physical properties and system design – CF31 extinguishant
158	15835-2009	Gaseous fire extinguishing systems – HCFC 125 extinguishing systems
159	15908-2011	Code of practice for selection, installation and maintenance of control and indicating equipments for fire detection and alarm systems
160	16018-2012	Specification for wheeled fire extinguisher – Performance and construction
161	16088-2012	Specification for chlorinated polyvinyl chloride (CPVC) pipes for automatic sprinkler fire extinguishing systems
162	SP 7-2005	National Building Code of India, 2016
163	SP 30-2011	National Electrical Code, 2011

A3.3 OIL INDUSTRY SAFETY DIRECTORATE

Sl. No.	Standard	Name of Standard
1	OISD-STD-105	Work permit system
2	OISD-RP-110	Recommended practices on static electricity
3	OISD-STD-113	Classification of area for electrical installations at hydrocarbon processing and handling facilities
4	OISD-GDN-115	Guidelines on fire-fighting equipment and appliances in petroleum industry
5	OISD-STD-116	Fire protection facilities for petroleum refineries and oil/gas processing plants

(Contd.)

6	OISD-STD-117	Fire protection facilities for petroleum depots, terminals, pipeline installations and lube oil installations
7	OISD-STD-118	Layouts for oil and gas installations
8	OISD-STD-142	Inspection of fire fighting equipments and systems
9	OISD-STD-144	Liquefied petroleum gas (LPG) installations
10	OISD-GDN-145	Guidelines on internal safety audits (procedures and checklist)
11	OISD-RP-149	Design aspects for safety in electrical systems
12	OISD-STD-150	Design and safety requirements for liquefied petroleum gas mounded storage facility
13	OISD-STD-156	Fire protection facilities for ports handling hydrocarbons
14	OISD-STD-164	Fire proofing in oil and gas industry
15	OISD-GDN-168	Emergency preparedness plan for marketing locations of oil industry
16	OISD-GDN-169	Guidelines on small LPG bottling plants (design and fire protection facilities)
17	OISD-STD-173	Fire prevention and protection system for electrical installations
18	OISD-STD-176	Safety health and environment training for exploration and production (upstream) personnel
19	OISD-GDN-180	Lightning protection
20	OISD-STD-189	Standard on fire-fighting equipment for drilling rigs, work over rigs and production installations
21	OISD-STD-191	Oil field explosive safety
22	OISD-GDN-192	Safety practices during construction
23	OISD-GDN-206	Guidelines on safety management system in petroleum industry
24	OISD-GDN-211	Safety in petroleum laboratories
25	OISD-GDN-227	Emergency response and preparedness in E&P industry
26	OISD-GDN-228	Selection, training and placement of fire operators in hydrocarbon and petrochemical industry

Appendix 4

Fire Properties of Some Flammable Materials

	Chemical formula	*Molecular mass*	*LFL*	*UFL*	*Flash point (°C)*	*Ignition point (°C)*	*IEC Gas Group*	*MIE (mJ)*
				Pure Compounds				
Hydrogen	H_2	2	4	75	–	500	IIC	0.011
Methane	CH_4	16	5	15	–	580	IIA	0.3
Acetylene	C_2H_2	26	2.5	81	–	305	IIC	0.017
n-Propane	C_3H_8	44	2	10	–	470	IIA	0.26
n-Butane	C_4H_{10}	58	1.9	8.4	–	405	IIA	0.26
Cyclohexane	C_6H_{12}	84	1.3	8	(–) 18	465	IIA	0.22
n-Hexane	C_6H_{14}	86	1.3	7	(–) 22	225	IIA	0.29
Benzene	C_6H_6	78	1.35	6.65	(–) 11	560	IIA	0.22
Toluene	C_7H_8	96	1.3	6.8	5	530	IIA	0.24
n-Heptane	C_7H_{16}	100	1	6	(–) 4	215	IIA	0.24
p-Xylene	C_8H_{10}	106	1	7	27	528	IIA	0.2
Iso-octane	C_8H_{18}		0.8	6	(–) 12	447	IIA	1.35
Methanol	CH_3OH	32	6.7	36	11	470	IIA	0.14
Ethanol	C_2H_5OH	46	3.3	19	13	365	IIA	0.65
Isopropyl alcohol	C_3H_7OH	60	2	13	12	425	IIA	0.65
Cyclohexanol	$C_6H_{11}OH$	100	1	9	67	300	IIA	
Acetone	$(CH_3)_2CO$	58	2.6	12.8	(–) 18	465	IIA	1.15
Ammonia	NH_3	17	15	28	–	630	IIA	680
Hydrogen sulphide	H_2S	34	4.3	46	–	260	IIB	0.07
Carbon monoxide	CO	28	12	75	–	609	IIA	<0.3
Carbon disulphide	CS_2	66	1.3	50	(–) 22	90	IIC	0.009

(Contd.)

	Chemical formula	*Molecular mass*	*LFL*	*UFL*	*Flash point (°C)*	*Ignition point (°C)*	*IEC Gas Group*	*MIE (mJ)*
				Common Fuels				
Aviation kerosene	–	–					IIA	0.2
Superior kerosene	–	–	0.7	5	35 (Min)	295	IIA	
High-speed diesel	–	–	0.6	6	32–96	225	IIA	

LFL	Lower flammability limit (% in air)
UFL	Upper flammability limit (% in air)
MIE	Minimum ignition energy

Note: Data from various sources. There may be variations on data depending on source.

Suggested Further Reading

Accidental Deaths and Suicides in India, National Crime Records Bureau, Ministry of Home Affairs, Government of India, Website: http://ncrb.gov.in.

Admirall, J.R. and Furton, K.G., *Analysis and Interpretation of Fire Scene Evidence*, CRC Press, USA.

Approved Document B (Vols. 1 and 2), Department for Communities and Local Government and Building Regulations Advisory Committee, UK.

Cooke, R.A. and Ide, R.F. (1985), *Principles of Fire Investigation*, The Institution of Fire Engineers, UK.

De Haan, J.D. and Icove, D.J., *Kirk's Fire Investigation*, Pearson Education Inc., New Jersey.

Drysdale, D.D., *An Introduction to Fire Dynamics*, Wiley-Blackwell.

Emergency Procedures Guidance, Health and Safety Executive, UK.

Essentials of Fire Fighting, International Fire Service Training Association (IFSTA), Oklahoma State University, USA.

Fire and Arson Scene Evidence, Technical Working Group on Fire/Arson Scene Investigation, National Institute of Justice, USA.

Fire Protection Handbook (2008), National Fire Protection Association, USA.

Fire Service Manual, Stationary Office, HM Fire Service Inspectorate Publications Section, UK.

Major Hazard Control — A Practical Manual (1988), International Labour Office, Geneva, Switzerland.

Mannan, S. (Ed.), *Lee's Loss Prevention in the Process Industries,* 4th ed., Elsevier.

National Bulding Code of India, Bureau of Indian Standards.

NFPA 1600, Standard on Disaster/Emergency Management and Business Continuity Programs, National Fire Protection Association, USA.

Oil Industry Safety Directorate (OISD) Standards, Ministry of Petroleum & Natural Gas, Govt. of India.

Quintiere, J.G., *Fundamentals of Fire Phenomena*, John Wiley & Sons, Sussex, UK.

Quintiere, J.G., *Principles of Fire Behaviour*, Delmar.

Relevant NFPA Standards, National Fire Protection Association, USA.

SFPE Handbook of Fire Protection Engineering, 5th ed. (2016), Society of Fire. Protection Engineers, USA and National Fire Protection Association, USA.

The *CIA Emergency Procedures Guide* (1976), Chemical Industries Association, London, UK.

World Fire Statistics 29 (2014), The Geneva Association, International Association for the Study of Insurance Economics, Geneva, Switzerland.

Zalosh, R.G., *Industrial Fire Protection*, John Wiley & Sons, Sussex, UK.

Index